Empowering Professional Teaching in Engineering

Sustaining the Scholarship of Teaching

Synthesis Lectures on Engineering

Each book in the series is written by a well known expert in the field. Most titles cover subjects such as professional development, education, and study skills, as well as basic introductory undergraduate material and other topics appropriate for a broader and less technical audience. In addition, the series includes several titles written on very specific topics not covered elsewhere in the Synthesis Digital Library.

Empowering Professional Teaching in Engineering: Sustaining the Scholarship of Teaching
John Heywood
2018

The Human Side of Engineering
John Heywood
2017

Geometric Programming for Design Equation Development and Cost/Profit Optimizaton, Third Edition
Robert C. Creese
2016

Engineering Principles in Everyday Life for Non-Engineers
Saeed Benjamin Niku
2016

A, B, See... in 3D: A Workbook to Improve 3-D Visualization Skills
Dan G. Dimitriu
2015

The Captains of Energy: Systems Dynamics from an Energy Perspective
Vincent C. Prantil and Timothy Decker
2015

Lying by Approximation: The Truth about Finite Element Analysis
Vincent C. Prantil, Christopher Papadopoulos, and Paul D. Gessler
2013

Simplified Models for Assessing Heat and Mass Transfer in Evaporative Towers
Alessandra De Angelis, Onorio Saro, Giulio Lorenzini, Stefano D'Elia, and Marco Medici
2013

The Engineering Design Challenge: A Creative Process
Charles W. Dolan
2013

The Making of Green Engineers: Sustainable Development and the Hybrid Imagination
Andrew Jamison
2013

Crafting Your Research Future: A Guide to Successful Master's and Ph.D. Degrees in Science & Engineering
Charles X. Ling and Qiang Yang
2012

Fundamentals of Engineering Economics and Decision Analysis
David L. Whitman and Ronald E. Terry
2012

A Little Book on Teaching: A Beginner's Guide for Educators of Engineering and Applied Science
Steven F. Barrett
2012

Engineering Thermodynamics and 21st Century Energy Problems: A Textbook Companion for Student Engagement
Donna Riley
2011

MATLAB for Engineering and the Life Sciences
Joseph V. Tranquillo
2011

Systems Engineering: Building Successful Systems
Howard Eisner
2011

Fin Shape Thermal Optimization Using Bejan's Constructal Theory
Giulio Lorenzini, Simone Moretti, and Alessandra Conti
2011

Geometric Programming for Design and Cost Optimization (with illustrative case study problems and solutions), Second Edition
Robert C. Creese
2010

Survive and Thrive: A Guide for Untenured Faculty
Wendy C. Crone
2010

Geometric Programming for Design and Cost Optimization (with Illustrative Case Study Problems and Solutions)
Robert C. Creese
2009

Style and Ethics of Communication in Science and Engineering
Jay D. Humphrey and Jeffrey W. Holmes
2008

Introduction to Engineering: A Starter's Guide with Hands-On Analog Multimedia Explorations
Lina J. Karam and Naji Mounsef
2008

Introduction to Engineering: A Starter's Guide with Hands-On Digital Multimedia and Robotics Explorations
Lina J. Karam and Naji Mounsef
2008

CAD/CAM of Sculptured Surfaces on Multi-Axis NC Machine: The DG/K-Based Approach
Stephen P. Radzevich
2008

Tensor Properties of Solids, Part Two: Transport Properties of Solids
Richard F. Tinder
2007

Tensor Properties of Solids, Part One: Equilibrium Tensor Properties of Solids
Richard F. Tinder
2007

Essentials of Applied Mathematics for Scientists and Engineers
Robert G. Watts
2007

Project Management for Engineering Design
Charles Lessard and Joseph Lessard
2007

Relativistic Flight Mechanics and Space Travel
Richard F. Tinder
2006

Empowering Professional Teaching in Engineering: Sustaining the Scholarship of Teaching

John Heywood

ISBN: 978-3-031-79381-3   paperback
ISBN: 978-3-031-79382-0   ebook
ISBN: 978-3-031-79383-7   hardcover

DOI 10.1007/978-3-031-79382-0

A Publication in the Springer series
SYNTHESIS LECTURES ON ENGINEERING

Series ISSN
Print 1939-5221   Electronic 1939-523X

Empowering Professional Teaching in Engineering

Sustaining the Scholarship of Teaching

John Heywood

Trinity College Dublin-University of Dublin

Foreword by Arnold Pears
KTH Royal Institute of Technology

SYNTHESIS LECTURES ON ENGINEERING #29

ABSTRACT

Each one of us has views about education, how discipline should function, how individuals learn, how they should be motivated, what intelligence is, and the structures (content and subjects) of the curriculum. Perhaps the most important beliefs that (beginning) teachers bring with them are their notions about what constitutes "good teaching". The scholarship of teaching requires that (beginning) teachers should examine (evaluate) these views in the light of knowledge currently available about the curriculum and instruction, and decide their future actions on the basis of that analysis. Such evaluations are best undertaken when classrooms are treated as laboratories of inquiry (research) where teachers establish what works best for them.

Two instructor centred and two learner centred philosophies of knowledge, curriculum and instruction are used to discern the fundamental (basic) questions that engineering educators should answer in respect of their own beliefs and practice. They point to a series of classroom activities that will enable them to challenge their own beliefs, and at the same time affirm, develop, or change their philosophies of knowledge, curriculum and instruction.

KEYWORDS

accountability, action research, active learning, advanced organiser, affective, animation, answerability, assessment, attitudes, beginning engineering educators, code of ethics, cognitive dissonance, communication, community, competence, complexity, cognitive organisation, curriculum (design, paradigms, process), concept (cartoons, clusters, inventories, key, maps, learning), content (syllabus), convergent, creativity, critical thinking, debates, decision making, design, diagnosis, discipline (s) (of knowledge), discovery, divergent, educational connoisseurship, evaluation, examinations (tests) ,examples, experts, expository instruction, instructional design, expressive activities, grading, heuristic(s), guided design, inquiry based learning, instructor centred, intellectual development, intelligence (applied, emotional, practical, academic), interdisciplinary, kinesthetic activities, knowledge (fields of, forms of, prior procedural, tacit, knowing), laboratory work, language(s), learner, learner centred, learning (active, independent, modes of, perceptual, surface, deep, styles of), lesson planning, lectures, listening, mediating response, memory, mind maps, misperception, mock trials, motivation, negotiate(ion),novice(s), objectives (behavioral/focussing), originality, outcomes, principles, professionalism (restricted/extended), reflection, Reflective Judgment Interview, peer teaching/review, personality types, philosophies related to engineering education, Polya, practical reflection, qualitative thinking, questions, questioning, scholar academic ideology, scholarship of teaching, social efficiency ideology, social reconstruction ideology, stages of development, taxonomies, teaching as research, tests, testing

Contents

Foreword xiii

Preface and Introduction xv

Acknowledgments xxi

1 Accountable to Whom? Learning from Beginning Schoolteachers 1 1

   1.1   Introduction 1

   1.2   Accountability in Higher and Engineering Education 1

   1.3   Accountability and Evaluation in Schools 2

   1.4   Accountability and Professionalism 3

        Notes and References 7

2 "Oh that we the gift of God to see ourselves as others see us," Learning from Beginning Teachers 2 11

   2.1   Introduction 11

   2.2   Recording One's Class 11

   2.3   Perceptual Learning in the Classroom 12

   2.4   Elliot Eisner's Concept of Educational Connoisseurship 13

        Notes and References 17

   2.5   Appendix 20

3 Toward a Scholarship of Teaching. Teaching as Research 25

   3.1   Introduction 25

   3.2   The Scholarship of Teaching 26

   3.3   Teaching and Design 31

   3.4   Teaching as Research–An Approach to Scholarship 33

        Notes and References 34

   3.5   Appendix 35

4 Objectives and Outcomes **39**

   4.1   The Social Efficiency Ideology 39

   4.2   The Objectives Movement 39

   4.3   The Taxonomy of Educational Objectives 40

   4.4   Eisner's Objections to the Objectives Approach 43

   4.5   Instructional Planning 47

   4.6   Questioning, Questions, and Classroom Management 48

   4.7   Reconciliation: A Conclusion 50

        Notes and References 51

5 Problem Solving, Its Teaching, and the Curriculum Process **57**

   5.1   Introduction 57

   5.2   Definitions and Approaches to Teaching Problem Solving 59

   5.3   Types of Problem, Difficulty, and Complexity 60

   5.4   Assessment, Instruction, and Objectives–The Curriculum Process 62

   5.5   Difficulty in, and Time for Learning 65

        Notes and References 66

6 Critical Thinking, Decision Making, and Problem Solving **71**

   6.1   Introduction 71

   6.2   Teaching a Decision Making Heuristic 72

   6.3   Qualitative Strategies 73

   6.4   Critical Thinking 76

   6.5   A category for Problem Solving? 77

   6.6   Looking Back Over Journeys 4, 5, and 6 79

        Notes and References 80

7 The Scholar Academic Ideology of the Disciplines **87**

   7.1   Introduction 87

   7.2   The Received Curriculum or the Scholar Academic Ideology 87

   7.3   The Post Sputnik Reform Projects 89

   7.4   Discovery (inquiry) Based Learning 90

   7.5   Is Engineering a Discipline? 96

        Notes and References 96

8   Intellectual Development 109
   8.1   The Spiral Curriculum 109
   8.2   Engineering and the School Curriculum 110
   8.3   Curriculum Questions Raised by Piaget's Theory of Cognitive Development 114
   8.4   Intellectual Development: Perry and King and Kitchener 114
       Notes and References 118

9   Organization for Learning 123
   9.1   Introduction 123
   9.2   The "Advanced Organizer" 123
   9.3   Using "Advanced Organizers" 124
   9.4   Prior Knowledge; Memory 125
   9.5   Cognitive Organization 125
   9.6   Mediating Responses 126
   9.7   Impact of K-12 and Career Pathways 127
       Notes and References 128

10   Concept Learning 131
   10.1   Robert Gagné 131
   10.2   Misperceptions 133
   10.3   Using Examples 134
       Notes and References 137

11   Complex Concepts 143
   11.1   Complex and Fuzzy Concepts 143
   11.2   Staged Development 144
   11.3   Concept Mapping and Key Concepts 145
       Notes and References 149

12   The Learning Centered Ideology–How Much Should We Know About Our Students? 153
   12.1   Introduction 153
   12.2   Communities of Practice, Communities that Care 154
   12.3   Learning Styles 155
   12.4   Convergent and Divergent Thinking 156
   12.5   Kolb's Theory of Experiential Learning 157

12.6   Felder-Solomon Index of Learning Styles 161
12.7   Temperament and Learning Styles 162
        Notes and References 164

13   Intelligence **173**
13.1   IQ and its Impact 173
13.2   Psychometric Testing 174
13.3   Controversies 175
        Notes and References 176

14   Two Views of Competency **181**
14.1   Nature vs. Nurture: Nature and Nurture 181
14.2   Inside and Outside Competencies 182
        Notes and References 183

15   From IQ to Emotional IQ **185**
15.1   Introduction 185
15.2   Implicit Theories of Intelligence, Formal, and Unintended but Supportive . . 186
15.3   Emotional Intelligence 193
15.4   Practical Intelligence 195
        Notes and References 198

16   Social Reconstruction **203**
16.1   The Fourth Ideology 203
16.2   Constructive Controversy 204
16.3   Debates 204
16.4   Mock Trials 205
16.5   Turning the World Upside Down 206
16.6   A case Study for Conclusion 206
        Notes and References 207

Author's Biography **209**

Author Index **211**

Subject Index **219**

Foreword

Tertiary education has experienced both rapid evolution and several significant changes in mission since the Second World War. Much of the technologically advanced world has become increasingly reliant on tertiary education as a supplier of engineers and creative thinkers of all types. At the same time, this utilitarian view of education has transformed the public view of education, which more often than not these days is seen as a process through which graduates are "produced", or as a "service" provided to an intellectual elite, which equips them for a successful and highly paid career. The view that education is about developing the individual and enhancing their intellectual capacity in the context of an academic environment which stimulated debate and enquiry has largely fallen by the wayside.

In this new landscape academic teachers are expected to perform research and teaching of the highest quality. High expectations in regard to teaching excellence has ben increasingly emphasised in the Nordic Countries, where in many places ten full time weeks of formal training in the theory and practice of tertiary education is a prerequisite for appointment to a tenure track position. Even in the United States of America the expectations in regard to teaching have changed significantly, not least in response to Boyer's 1991 book "Scholarship Reconsidered: Priorities of the Professoriate".

Quality in higher education is also an increasingly prominent component of the political discourse surrounding tertiary education. This book makes a significant contribution to both academic staff development and teaching quality by drawing together over fifty years of work in the area of evidence based teaching practice. The reader gains both new perspectives on teaching and assessment practices and a model for sustainable practice and professional development as a university teacher. Academic practice is more than research, the educational mission to inspire future generations of scholars to engagement and excellence in science and engineering underpins the success of our technological society.

The model and resources offered here form part of a broader effort in which Professor Heywood, myself, the American Society for Engineering Education (ASEE) and the IEEE Education Society are engaged. The goal is to provide sustainable support for academic teaching practice and professional development combined with international levels of professional recognition linked to a range of activities that promote and enhance the "Teaching as Research" model. This book is a vital resource in the pursuit of this goal, and it gives me great pleasure to have contributed in a small way to its conception and final form.

Arnold Pears
Professor and Chair of the Department of Learning in Engineering Sciences
KTH Royal Institute of Technology
Stockholm, Sweden
July 2017

Preface and Introduction

At the 2016 ASEE/IEEE Frontiers in Education Conference (FIE) Professor Arnold Pears of Uppsala University in Sweden organized and led a one day workshop on teaching and assessment for beginning engineering educators and experienced engineering educators beginning to take an interest in teaching. I was privileged to lead the discussion on assessment. I noticed that several of the participants were experiencing the same difficulties that beginning school teachers experienced, and drafted some notes that I thought might be used in any future courses of this kind. Dr Mani Mina of Iowa State University with whom I had collaborated in presenting a blended on line course on, "The Human Side of Engineering" attended the workshop, and as a result of my notes it was decided that he would organize a professional development course on teaching and learning for his colleagues in the Departments of Electrical and Computer Engineering and Industrial Design. It would use the format of the previous course. In the event 16 lectures each of approximately 20 minutes duration were recorded, and followed four days later by hour long discussion seminars on the prior recorded topic. A print version was also made available. This book records the sixteen lectures with the associated notes which are of equal importance.

The first three journeys are constructed around the issue of accountability. To whom am I accountable, and for what? Many engineering educators experience a conflict between the demands of research and the requirements for teaching. Looked at from the perspective of professionalism, a person who enters engineering education acquires a dual responsibility for research and teaching. Irrespective of the demands for and recognition achieved by research, there is an obligation to be as effective as possible at teaching. By accepting the role of engineering educator an individual accepts that teaching is a professional activity, and has to choose between being a "restricted" or an "extended" professional. Professionals accept personal responsibility for the effectiveness of their teaching. How individuals can judge the effectiveness of their teaching is the subject of journeys and two and three. Journey 2 focuses on Eisner's technique of educational connoisseurship, and Journey 3 considers what the scholarship of teaching is, and argues that it is accomplished by treating the classroom as a laboratory for research and development. Effective teaching can only be sustained if that becomes the case. This requires an acknowledgement and understanding of that body of knowledge called "education." This book is one way of introducing that body of knowledge.

Each one of us has views about education, how discipline should function, how individuals learn, how they should be motivated, what intelligence is, and the structures (content and subjects) of the curriculum. Perhaps the most important belief that beginning teachers bring with them are their beliefs about what constitutes "good teaching". The scholarship of teaching requires that beginning teachers should examine these views in the light of knowledge currently available about the curriculum and instruction.

Since there is no single theory of the curriculum or instruction various attempts have been made to classify the different ideologies that represent the diversity of views among engineering educators.. In Britain John Eggleston distinguished between "received", "reflexive", and "restructuring" paradigms of the curriculum. In the United States Michael Schiro distinguished between four ideologies that he called "Scholar Academic", "Social Efficiency", "Centred", and "Social reconstruction". The philosophies that support these ideologies also support different approaches to teaching. Michael Schiro reports one research that shows that teachers change their beliefs during their teaching careers.

Journey 4 begins with the social efficiency ideology for the reason that it is this ideology that governs much educational thinking at the present time, and in engineering in particular. It begins with a brief account of the "objectives" movement leading to a discussion of the "Taxonomy of Educational Objectives", and objections to the objectives approach by Eisner. The role of objectives in planning and instruction is considered. The journey ends with an attempt to reconcile the behavioral objectives approach with that of its opponents.

The fifth Journey considers the problem of problem solving. Should it be taught as a separate skill or simply learnt by total immersion in the subject? Those who hold the former view are representative of the social efficiency ideology. A distinction may be made between those who believe problem solving should be taught within normal course structures, and those who believe it should be taught in separate courses. The best known example of the latter is the Problem Based Learning approach developed by Don Woods at McMaster University. There are many examples of the former where teachers use a simple problem solving heuristic like that suggested by Polya as part of their instructional approach. It is with this approach that Journey 5 is primarily concerned. It shows just how difficult the curriculum process is, and how "time" is required for learning.

Journey 6 is a continuation of Journey 5 and looks at problem solving heuristics in more detail, and in particular at Wales, Stager and Nardi's "Guided Design" model. Studies of expert and novice behavior reported in Journey 5 and this journey, showed there was something more to problem solving in engineering than the learning of a range of heuristics, and that there was a need for qualitative as well as quantitative understanding. Engineers have to learn a number of languages if they are to successfully engage in engineering problem solving. It is concluded that there is a case for a separate category of problem solving in any statement of objectives.

These three Journeys (4, 5, 6) highlighted the importance of assessment on learning. They showed how changing the conditions of learning impact on the role of the teacher. They also pointed to questions about students. What should instructors know about their students? How do teacher beliefs impact on what they do? For many teachers these beliefs may be described as belonging to the scholar academic ideology, or Eggleston's received curriculum.

Journey 7 introduces the scholar academic ideology. In a received curriculum knowledge is received and accepted as given. It is non-negotiable, non-dialectic, and consensual. It is the basis of the "disciplines" view of the curriculum. It is about the enculturation of individuals into civilization's accumulated knowledge and ways of knowing. But, each discipline seeks to mould students in its own image and likeness. Many academics including engineering educators would associate themselves with this ideology. It is teacher centred. Jerome Bruner who is associated with this ideology is of particular interest because of his promotion of discovery (now often called inquiry) based learning. The advantages and disadvantages of this kind of learning are considered. The journey ends with a brief section headed by the question "Is engineering a discipline?"

Associated with Jerome Bruner is the idea of "spiral curriculum" in which concepts are revisited on several occasions during the course, but at deeper levels of abstraction. Journey 8 begins with a discussion of this model. It raises questions about how engineering is related to the school curriculum, and an example of a primary school project in which children in the age range 5 to 13 engaged in min-company activities is given Those who sponsored the activity believed that entrepreneurs would only emerge if attention was paid to the development of entrepreneurial skills throughout the age range of schooling. The Spiral curriculum also relates to intellectual development. The significance of Piaget's work, and studies of intellectual development in higher education by Perry, and King and Kitchener are considered.

Bruner's discovery learning was criticised by among others David Ausubel. Although a very strong advocate of expository learning, he was concerned with the way in which learning is organized. He is noted for the concept of the "advanced organizer". Its use in instructional practice begins Journey 9. The importance of prior knowledge in learning and the development of memory is emphasized. The journey ends with a discussion or cognitive organization and mediating responses. Much care needs to be taken in the preface to instruction if that instruction is to be meaningful to students

Meaningful learning requires that students understand concepts. The role of concepts in learning, and in particular the work of Robert Gagné is the subject matter of Journeys 10 and 11. One of the reasons why students find qualitative thinking in engineering difficult is that they have an inadequate understanding of concepts to the extent that they are misperceived. How to deal with misconceptions is a major problem for instructors. The most common heuristic used in instruction is the "example". Research shows that some approaches to the use of examples are better than others. Learning concepts often takes time and many teachers do not take a step by step approach because of beliefs about the need to cover the syllabus. This seems to be a central issue in teaching. It seems probable that a lot of the difficulties experienced by engineering students, especially in the freshmen year, arise from a shortage of time to assimilate the learning of the concepts being presented especially when they are complex. Journey 11 gives a brief introduction to the teaching of complex and fuzzy concepts.

The focus of Journey 12 is on the learner centred ideology. It is in stark contrast to the social efficiency ideology. The child is at the centre of, and has a profound influence on the curriculum process. Like the social reconstruction ideology it is associated with the philosophy of John Dewey. In this ideology the student is a self-activated maker of meaning. Learning moves from the concrete to the abstract. Learning centred educators know a lot about their students. It is argued that engineering educators should have at least a knowledge of their students learning styles. The journey draws attention to convergent and divergent thinking because there is strong argument that engineering students are often taught in ways that are antipathetic to creative thinking. Following discussion of Kolb's theory of experiential learning and the Felder-Solomon Index of Learning Styles, the journey concludes with a brief commentary on the relation between temperament and learning styles. It is concluded that studies of learning styles and the temperaments of students can provide educators with insights into student learning and instruction.

Those who follow the learning centred ideology do not like psychometric testing or formal examinations. Yet most of us have beliefs about intelligence and its role in learning. Journeys 13, 14, and 15 deal with issues surrounding the concept of intelligence. Journey 13 begins with a brief discussion of the impact that intelligence testing has had on school systems. It is agreed that tests of general mental ability are found to be relatively good predictors of job performance. But multiple methods of assessment are to be preferred to a unitary instrument. Journey 14 begins with a description of the nature-nurture controversy and concludes that we should think about "Nature and Nurture" not "Nature versus Nurture".

Just as engineering educators should have a view about intelligence so they should have a view about competence. Two views of competence are presented. They have profound consequences for the design of the curriculum and instruction. The role of communication is highlighted, but doubt is cast on the methods used to teach communication as a means of achieving the goals that are required. The view is expressed that the curriculum should be perceived in terms of intellectual and personal development that continues throughout life. That places considerable responsibility on industry for the development of their personnel which most organizations do not seem to accept.

Two alternative theories of intelligence are presented in Journey 15. The first is Howard Gardener's theory of multiple intelligences, and the second, Robert Sternberg's Triarchic Theory of intelligence. Attention is given to implicit theories of intelligence. Sternberg is also important for engineering education because of his concept of "practical intelligence." The journey ends with a discussion of emotional intelligence. These journeys show that not only teaching but policy making in respect of the curriculum, benefit if we have a wide ranging understanding of student behavior.

The final journey is a commentary on the social reconstruction ideology. It considers that society is doomed because its institutions are incapable of solving the social problems with which it is faced. Therefore, education has to concern it with the reconstruction of society. Like the learning centred ideology it is based on a social constructivist view of knowledge. The principle methods of teaching are "discussion" and "experience" group methods. In education Karl Smith has encouraged "constructive controversy". Other methods are "debates" and "mock trials". The journey ends with a case study. It is concluded that since learning is shared activity the least an instructor can do to foster relationships is to share his/her scholarly activity with his/her students.

John Heywood
October 2017

Acknowledgments

I am very grateful to Professor Arnold Pears of Uppsala University for inviting me to participate in this project which I have enjoyed immensely.

A big thank you to Dr Mani Mina of Iowa State University for organising this lecture programme and for being my critical friend.

He and I would like to thank Farah Nordin for the large amount of time she gave to the project to tape, and edit the video and audio files. We would also like to thank Mr Kevin Wikham of the Department of Electrical and Computer Engineering for helping with the web development and WordPress set.

We would like to thank Professor David Ringholz, Professor Steve Herrnstad, Matthew Krise, Peter Evens and the faculty, graduate and undergraduate students of the Department of Industrial Design for their continuing interest and enthusiastic support for the project.

More especially we would like to thank the following for leading and contributing to the seminar discussions – Neelam Prabhu-Gaukar, Sara jones, Leif Buaer, Mohammed Al-Mokhainin, and Professors John Basard and Lofthi Ben-Otheman.

John Heywood
October 2017

J O U R N E Y   1

Accountable to Whom? Learning from Beginning Schoolteachers 1

1.1   INTRODUCTION

The engineering profession has been keen to develop engineering activities in schools. Both the ASEE and FIE annual conferences hold several sessions each year devoted to K-12 education in which there are exchanges about what has been done and what might be done. Occasionally it is pointed out that engineering education can learn to its advantage about teaching methods in schools especially in primary (elementary) schools [1, 2]. There are no detailed analyses of engineering educators at work of the kind carried out among school teachers by Lortie [3] and more or less replicated twenty years later by Cohn and Kottkamp in the United States [4].

My experience of teacher education leads me to believe that beginning engineering educators have much to learn from beginning teachers. Therefore, many examples in this text are taken from reports of what happened to beginning teachers and their students while researching their own instruction.

There seems to be general agreement that there is a need for induction to teaching that goes beyond telling beginning teachers where their classroom, rest rooms, and staff rooms are before they begin their teacher training. However, by all accounts engineering education is still at this primitive stage. It is not unreasonable to suppose that key questions on a beginning engineering educators mind relate to accountability: "to whom, and for whom am I responsible?"

1.2   ACCOUNTABILITY IN HIGHER AND ENGINEERING EDUCATION

Accountability is important because it is the devil that is driving the mechanisms that control the work of teaching, as for example, the ABET criteria. In the UK, higher education institutions are now being judged for their teaching quality as well as their research [5]. That is, in addition to the quality assurance procedures already in place.

To begin at the beginning, Sockett wrote in 1980 that: "Central to the debate on accountability are the twin ideas of responsibility and answerability for actions undertaken by one party on behalf of another" [6]. My version of the development of accountability in the education system in England is that it began with the student revolt of 1969. Parliamentarians found that although the student unions in the universities received funding from student fees they were not required to account for how it was spent, and this frustrated those parliamentarians. They also came to believe that the measures in place for checking on the usage of funds by the universities were not adequate. In consequence, and it took a long time, the funding and accountability mechanisms were changed in the latter half of the nineteen-eighties. By far the most important control mechanism became the regular review of research, that is, the rating of departments against the number of publications produced, their quality as measured by peer review and the medium in which they were presented (e.g., conference, journal). Publications also became important in the United States for academics seeking tenure or promotion to associate and full professorship. Research became more important than teaching in many institutions, and the term research university coined.

Many beginning engineering educators are brought up in this system without any break in industry, and understand they have to publish or perish. The key qualification for progress into engineering education is the Ph.D., not paradoxically a Ph.D. and a qualification to teach. Some beginners may have had the experience of being a teaching assistant but few will have had any training for teaching, although training courses are available in some countries and compulsory in others [7]. A beginning teacher coming from industry will be in this situation but will have experienced the discipline of working in industry, and some of them find the organization and attitudes of engineering educators surprising. In either case both types are suddenly faced with role conflict between the relative efforts they should put into teaching on the one hand, and on the other hand, research. They come to a crossroads one of which points to research, and the other in the opposite direction toward teaching. As things stand unless teaching is formally appraised they are more likely take the research option.

Demands for improvement in teaching may increase the tension between research and teaching [8], and if undertaken within existing procedures for quality assessment create tensions between traditional teaching and innovative teaching as Pears has demonstrated for engineering in Sweden [9]. A few of the graduate student teachers whose exemplars are given in these chapters found the tasks I asked them to do brought them into conflict with their Master teachers.

1.3   ACCOUNTABILITY AND EVALUATION IN SCHOOLS

In parallel with these developments schools were also subject to similar pressures. However, in the UK, university education departments and colleges of education ensured that there was a substantial debate about accountability which extended to teacher education courses. Sometimes, as in my case, it was linked to problems associated with evaluation, since evaluation is a form of accountability [10]. I focused on the relationship between accountability and professionalism and argued, following Elliott, that the first point in the chain of accountability was the teacher.

Elliot wrote in 1976 that, "If teacher education is to prepare students or experienced teachers for accountability then it must be concerned with developing their ability to reflect on classroom situations. By 'practical reflections' I mean reflection with a view to action. This involves identifying and diagnosing those practical problems which exist for the teacher in his situation and deciding on strategies for resolving them. The view of accountability which I have outlined, with its emphasis on the right of the teacher to evaluate his own moral agency, assumes that teachers are capable of identifying and diagnosing their practical problems with some degree of objectivity. It implies that the teacher is able to identify a discrepancy between what he in fact brings about in the classroom and his responsibilities to foster and prevent certain consequences. If he cannot do this he is unable to assess whether or not he is obliged to. I believe that being plunged into a context where outsiders evaluated their moral agency without this kind of developmental preparation would be self-defeating since the anxiety generated would render the achievement of an objective attitude at any of these levels extremely difficult" [11].

Successful accountability is more likely to be achieved when teachers take responsibility for their daily actions at what might be deemed to be the first level of accountability. The second level, which cannot be avoided, relates that accountability to the outside world through appraisal, that is, of objectives agreed between the teacher and the authorities (principals, parents, colleagues) to whom he is accountable. Thus, if teachers wished to consider themselves to be professionals then, in the first instance, they had to be self-accountable for the achievement of agreed goals. They had to be able to self-evaluate or as we would say today, self-assess.

1.4   ACCOUNTABILITY AND PROFESSIONALISM

In the traditional concept of a profession the professional person is self-employed. As such they are necessarily self-accountable for their work, and this impacts, or should impact, on the service they provide their clients. But this idea was challenged in the nineteenth century and persons who were employed came to be regarded as professionals. In the 19th Century the creation and development of the engineering institutions, the Institution of Civil Engineers in the UK in particular, led to the view that engineering was a profession. In the 20th Century, particularly in the latter half, many other groups sought recognition as a profession from society. Teachers belonged to this group, and they became recognized as such, as did many other groups [12].

Lest it be thought that this argument only applied to the UK, it should be noted that in the U.S. in 1970, Owens argued that since the teachers are professionals they should be responsible for what goes on in the classroom. The teacher is no different to the medical practitioner in this respect [13]. But those who teach in higher education do not regard teaching as a professional activity. In engineering their allegiance is to the engineering profession, and their research is associated with that allegiance. This is one of the, if not the major reason why faculty do not have much interest in aligning their teaching and assessment to the knowledge base of techniques that is available to them. It is one of the reasons why it is so difficult to reform or change the practices of engineering education.

The best that can be said of the majority of engineering educators is that they are "restricted" professionals to use a term coined by Eric Hoyle. He made a distinction between "restricted" and "extended" professionalism. He argued that at that time teachers looked for a restricted notion of professionalism which is "a high level of classroom competence teaching skill and good relationships with pupils" [14]. And this is what the public would expect. In Ireland, Henry Collins (sometime President of the Association of Secondary Teachers Ireland) examined Hoyle's model of restricted professionalism, and concluded that achieving the competence that the public expects of teachers would necessarily extend their professionalism [15].

"Extended professionalism" wrote Hoyle "embraces restricted professionalism, but additionally embraces other attitudes of the teacher. These include seeing his/her work in the wider context of community and society, ensuring that his/her work is informed by theory, research, and current exemplars of good practice; being willing to collaborate with other teachers in teaching, curriculum development and the formation of school policy, and having a commitment to keep himself/herself professionally informed."

Hoyle's model of restricted and extended professionalism is easily adapted for higher education as Exhibit 1.1 shows.

Engineering Educators who attend the annual ASEE and FIE conferences are more likely to be, or have a tendency toward extended professionalism, and to take the issue of accountability seriously.

An important step that would enable engineering educators to become a professional has been taken by the American Society for Engineering Education who have promoted a code of ethics. Cheville and Heywood have discussed the problems of developing a code of ethics for engineering education with reference to those in use in engineering (world-wide) and other professions, and suggested the code shown in Exhibit 1.2 [16]. As yet there is no universally recognized system of training engineering educators that is the hallmark of a profession.

It is evident from the foregoing that what has been written about accountability in the school system applies equally to higher education.

The beginning engineering educator finds her/himself in a situation where it seems that she/he may have to choose between different priorities or to balance those priorities as best as he/she can. In 1994 Michael Bassey President of the British Educational Research Association published a book with the title "*Creating Education through Research*" [17]. He used the second page of his introduction to adapt and develop work by W. G. Perry whose study of student intellectual development in higher education is of considerable importance [18, see chapter VIII]. The last two paragraphs read,

"My fifth discovery was that I am not a watcher of the world but an actor in it. I have to make decisions and some of them have to be made now. I cannot say, 'stop the world and let me get off for a bit, I want to think some more before I decide.' Given

Restricted Professionality in Engineering Education	Extended Professionality in Engineering Education
Instructional skills derived from experience	Instructional skills derived from mediation between experience and theory
Perspective limited to immediate time and place	Perspective embracing broader social context of education
Lecture room and laboratory events perceived in isolation	Lecture room and laboratory events perceived in relation to institution policies and goals
Introspective with regard to methods of instruction	Instructional methods compared with those of colleagues and with reports of practice
Value placed on autonomy in research and teaching	Value placed on professional collaboration in research and teaching
Limited involvement in non-teaching professional and collegial activities	High involvement in non-teaching professional and collegial activities
Infrequent reading of professional literature in educational theory and practice	Regular reading of professional literature in educational theory and practice
Involvement in continuing professional development limited and confined to practical courses mainly of a short duration	Involvement in continuing professional development work that includes substantial courses of a theoretical nature
Instruction (teaching) seen as an intuitive activity	Instruction (teaching) seen as a rational activity
Instruction (teaching) considered less important than research	Instruction (teaching) considered as important as research
Assessment is a routine matter. The responsibility for achievement lies with the student	Assessment is designed for learning Achievement is the co-responsibility of the institution, instructor (teacher) and student

Exhibit 1.1: An adaptation of Hoyle's descriptions of restricted and extended professionalism on school teaching for higher education.

Engineering education has a large impact on the world, serving the ideal of human development through education and the ideal of truth through scholarship. Engineering educators respect the impacts culture and individuality have on these ideals. To serve these ideals engineering educators:

1. recognize that engineers and engineering works may impact the world for good or for ill. Engineering educators strive to develop their own and students capacity for moral purpose, serve as an example for life lived well, and recognize the rights of others to define their own welfare and quality of life;

2. treat others fairly, support others' learning at all times, and honor differences between learners that arise through opportunity and culture;

3. balance responsibilities of the multiple roles they assume within the education system:

   a. in the role of a teacher or mentor the engineering educator seeks to support learning, professional development, and enabling human thriving through education;

   b. in the role of a scholar the engineering educator dedicates himself or herself to seeking truth and awareness of his/her own ideology;

   c. in the role of an administrator, the engineering educator is guided by principles of fairness, justice, and compromise;

   d. in the role of a patron, constituent, or client the engineering educator provides actionable feedback to improve education and helps support others professional development; and
   While most times these rules are harmonious, in some cases the engineering educator will face ethical dilemmas that arise from overlaps of these roles. Resolving such conflict requires both adherence to law and moral judgment, tempered with respect for colleagues and students, and the recognition that vulnerable populations may often lack a voice. The engineering educator acknowledges the tensions inherent in supporting individual learners and an educational system with limited resources while undertaking unbiased evaluation of learning.

4. serve educational needs through:

   a. supporting the needs of learners and upholding the rights of all individuals to an education with particular care for the vulnerable and disadvantaged;

   b. recognizing the impact of credentials and the limitations inherent to measuring learning, and striving to improve how learning is assessed;

   c. recognizing that learning occurs within a community and valuing the diverse expertise and contributions of their colleagues and the supports offered by the wider educational institution in which they function; and

   d. building professional liaisons with others across the education system, and those who employ engineering graduates.

5. uphold standards of professionalism in any role they play within the education system;

6. balance their role as an educator with their role as an engineer by accurately interpreting state-of-the-art engineering theory and practice for learners, and drawing upon the science of learning to effectively promote and support student development;

7. act in ways that develop and hold the trust and confidence of others so as to support their role as teacher and mentor;

8. seek to advance, apply, and integrate the state of the art in both education and engineering theory and practice and dedicate themselves to life-long professional development; and

9. recognize a responsibility to participate in activities that contribute to access to education, and seek changes to situations that are contrary to the best interests of learners.

Exhibit 1.2: A proposal for a code of ethics for engineering educators by R. A. Cheville and J. Heywood (Cheville, R. A. and J. Heywood (2015). Drafting a code of ethics for engineering education. *ASEE/IEEE Frontiers in Education Conference*, pp. 1420–1423).

differences of opinion among reasonable people, I realize that I cannot be sure that I am making the 'right' decisions. Yet because I am an actor in the world, I must decide. I must choose what I believe in and own the consequences."

This text is written for beginning engineering educators and engineering educators who have decided to give teaching the same value that they give to research, and to show how the exploration of techniques that have been used in the school system can help the development of skill in self-accountability, enable choices to be made about curriculum and instructional design, and thereby, to create education. Since we expect students to take responsibility for their learning, they in turn have the right to ask us to take responsibility for teaching excellence.

NOTES AND REFERENCES

[1] Crynes, B. L. and D. A. Crynes (1997). They already do it: authors]Crynes, B .L. authors]Crynes, D .A. Common practices in primary education that engineering education should use. *ASEE/IEEE Proceedings Frontiers in Education Conference*, 1219. 1

[2] Heywood, J. (2002). SCOOPE and other primary (elementary) school authors]Heywood, J. projects with a challenge for engineering education. *ASEE/IEEE Frontiers in Education Conference*, F2C-6 to 10. 1

[3] Lortie, D. C. (1975). *Schoolteacher. A Sociological authors]Lortie, D. C. Study*. Chicago, Chicago University Press. 1

[4] authors]Cohn, M. M. authors]Kottkamp, R. B. Cohn, M. M. and R. B. Kottkamp (1993). *Teachers. The Missing Voice of Education*. Albany, NY, State University of New York Press. 1

[5] authors]Hurst, G. Hurst, G. (2016). University teaching will be rated gold, silver and bronze. *The Times* p. 13, September 30. 1

[6] authors]Sockett, H. Sockett, H. (1980). *Accountability in the English Educational System*. London, Hodder and Stoughton. 2

[7] Training for university teachers. 2

Australia.   Most universities including the research universities insist that new staff complete at least a basic course on university teaching that includes peer evaluation of their teaching. Some encourage their staff to complete a graduate certificate of higher education. Most universities offer such a qualification and it is "free" to their staff. At The University of Technology Sydney new teaching academics are supposed to be given a reduced workload in their first year of employment to allow them to do the Grad Cert.

Sweden.   University teachers are now required to have pedagogic training. At Uppsala University there is a course specifically designed for engineering educators. The general requirement at Uppsala reads "a prerequisite of an applicant's educational competence as satisfactory are completed, relevant, pedagogy of higher education courses with a workload comprising a minimum ten full-time weeks, or equivalent knowledge. In special circumstances dispensation can be granted to allow the candidate to complete the required ten weeks of education during the first two years of employment. For appointments to professorships the pedagogical education must include a course in research supervision."

United Kingdom.   Initially intended to be compulsory but is now voluntary and overseen by the Higher Education Academy (formerly ILT) which offers Fellowships at four levels. It has established a UK Professional Standards Framework (UKPSF). Some universities offer post graduate certificates in higher education that are accredited by HEA. Many universities require new staff to pursue post graduate certificates in teaching and learning—usually two years part-time. Some universities require departments to have a proportion of qualified teachers. The new legislation may lead to this becoming a requirement.

United States.   Many universities in the United States have training programmes. These range from a few days like the NETI's (National Effective Teaching Institute) offered through the American Society for Engineering Education to more traditional structures of the kind offered at the University of Wisconsin-Madison where I was privileged to participate in a 2/3 credit course for graduate students on teaching Engineering and Science (See Courter Sandra and J. Heywood (2002). The perceptions of science and engineering graduate students to the educational theories relevant to skill development in curriculum leadership. *ASEE/IEEE Proceedings Frontiers in Education Conference*, F4A-1 to 5).

[8] William, D. (2016). Letter to *The Times* p. 20, September 30. authors]William, D. The British approach relies heavily on student ratings. For a critique of student ratings see Heywood, J. (2000). *Assessment in Higher Education. Student Learning, Teaching, Programmes and Institutions*. London, Kogan Page. Chapter 4. The assessment of teaching by students and alumni. 2

[9] Pears, A. N. (2009). Implications of student conceptions of authors]Pears, A. N. teaching for the reform of engineering education. *ASEE/IEEE Proceedings Frontiers in Education Conference*, WA-1 to 5. 2

[10] Heywood, J. (1984). *Considering the Curriculum authors]Heywood, J. during Student Teaching*. London, Kogan Page. 2

[11] Elliott, J. (1976). Preparing for classroom accountability. authors]Elliot, J. *Education for Teaching*, 100, pp. 49–71. 3

[12] Heywood, J. (1983). Professional studies and validation in C. authors]Heywood, J. H. Church (Ed.), *Practice and Perspective in Validation*. London. Society for Research into Higher Education. 3

[13] Owens, R. C. (1970). *Organizational Behaviour in authors]Owens, R. C. Schools*. Englewood Cliffs, NJ, Prentice Hall. 3

[14] Hoyle, E. (1973). Strategies of curriculum change in Watkins, authors]Hoyle, E. E. (Ed.), *In Service Training. Structure and Context*. London, Ward Lock. 4

[15] Collins, H. P. (1980). A study of some aspects of the status authors]Collins, H. P. of organized teachers within the education system. Med Thesis. Dublin. School of Education, University of Dublin. 4

[16] Cheville, R. A. and J. Heywood. (Cheville, R. A. and authors]Cheville, R. A. authors]Heywood, J. J. Heywood (2015). Drafting a code of ethics for engineering education. *ASEE/IEEE Frontiers in Education Conference*, pp. 1420–1423. 4

[17] Bassey, M. (1994). *Creating Education Through authors]Bassey, M. Research. A Global Perspective of Educational Research for the 21st Century*. Newark, Kirklington Moor Press in association with the British Educational Research Association. 4

[18] Perry, W. G. (1970). *Forms of Intellectual and authors]Perry, W. G. Ethical Development in College Years. A Scheme*. Troy, Mo. Holt Rinehart and Winston. 4

JOURNEY 2

"Oh that we the gift of God to see ourselves as others see us," Learning from Beginning Teachers 2

2.1   INTRODUCTION

I confess that the quotation is the only line of the poetry by Robert Burns that I know. At least I was told he wrote it, and I was also told that it was about a lice on the back of the neck of a Lady in her "Sunday best," who was, as one might expect, attending church.

Self-accountability demands that we try and see ourselves as others see us even if what we find is unpleasant. I take Rokeach's proposition that generally, we want to know the truth about ourselves, to be correct [1]. Engineering educators with an interest in teaching and learning are still a rare breed and finding a person or better still persons interested in teaching is often difficult yet, the height of self-accountability is to be able to invite a colleague to observe one's teaching. Such liaisons are the basis of educational change. You may read about an attempt by a primary (elementary) school teacher in Ireland to engage his colleagues in changing the curriculum of their school in appendix A (Section 2.5).

One important thing that a beginning teacher has to learn, if he or she does not know it already, is that students do not always see things in the same way as the teacher [2]. A component of the competency of self-accountability is to be able to see ourselves as others see us, more especially our students. The concern of this chapter is with techniques for achieving that goal.

2.2   RECORDING ONE'S CLASS

When John Elliott developed his thoughts about accountability in the 1970's, he suggested that the teacher should make an audio recording of his/her lesson/lecture [3]. At that time it would have been very expensive to have made a video recording and, in any event, the equipment would have been very large. Nowadays, we are probably being recorded by one or another of the students in our class! We can certainly prop a laptop with a camcorder on the lecture room's rostrum, whatever that may be, and make a visual recording of the activity.

I have nothing against that, but I want to suggest that from listening, yes- just by listening, we may learn a great deal about ourselves, the way we present knowledge, and the way we interact with students. It is much easier to do these days because we have the technology that makes omni-directional-recording easy.

Listening is an important skill and helps us to focus on the issue we want to study, as for example, how we respond to questions in class. It won't, of course, necessarily tell us if we are selective in the choices we make about whose question we will take, unless we do a more detailed analysis.

After a few audio sessions we can begin to make and analyze video recordings and cope with the much greater "noise" that is generated.

One of the other ways my colleagues used to train beginning teachers—called "microteaching"—was to bring in half-a-dozen students from a local school, ask the student to teach them for ten-minutes or so while making a video recording of their teaching. The recording is then played back to the student with comments from the tutor. This procedure can be changed so that a group of beginning teachers review their teaching together, and comment on each other's presentations. It is quite a useful method for introducing beginning teachers to the art of teaching. But, it is only the beginning of self-accountability.

The first time that I tried to make a videotape of an introductory lecture came as a shock. What I thought would be a doddle turned out to be very stressful. The producer harangued me and continually stopped to re-take and re-take. Apart from learning that it was a considerable skill I began to appreciate just how little a learner can address, that 50 minutes to an hour is far too long for a continuous presentation, and that it was very easy, even then with all the electronics available, to introduce noise into the learning system [4].

Farah and Neelam did their very best to make me presentable in the first of these mini-lectures. I had to do a lot of re-learning.

A quite different approach was advocated by the Stanford educator Elliot Eisner (see below).

2.3   PERCEPTUAL LEARNING IN THE CLASSROOM

One of the difficulties that student graduate teachers have in trying to understand classroom performance is to get behind (understand, if you prefer) what the students are thinking. In evaluating their classes the emphasis is often with what happens to them rather than what happens to their students as a result of their instruction. This is not at all surprising. At the same time it is a reminder that what happens to teachers in classrooms is all too easily forgotten by politicians and administrators when they criticize them. The teacher is as important as the student in the learning process, but the teacher has to be aware of the perceptual processes at work.

The relationship between teacher's and their students is deep and personal and can be encouraging or hurtful in both directions. That said, teachers do need to understand what is happening in their classrooms both to themselves and to their students.

The teachers who provided the examples given in this text taught in schools under supervision throughout the school year for half of each week. Throughout the academic year they attended the university during the other half of the week. At the beginning of the school year, to assist them with their practice, the department provided an induction course. The big worries that the student graduate teachers had were, would they be able to maintain discipline? And, would they be able to motivate their pupils? My particular contribution to this course was to introduce them to perception and perceptual learning. I wanted to achieve a number of objectives. Of these the one that is relevant to this text is that students do not always perceive what the teacher is saying as the teacher expects it to be perceived. Hence, the need for questioning and testing at the time to establish what is being learned and how it is understood [5]. The situation is no different in a university class, as Jane Abercrombie showed in studies of architectural and medical students. Her "*The Anatomy of Judgement*" first published in 1960 [6] must be a classic in the literature of higher education.

2.4   ELLIOT EISNER'S CONCEPT OF EDUCATIONAL CONNOISSEURSHIP

Although, in the 1970's Elliot Eisner wrote one of the most profound books on the curriculum he is seldom cited in the literature of engineering education. Possibly this was because he was anti-positivist, and was also a considerable critic of the objectives movement [7]. Eisner's world was that of art and design. From that world came the concept of connoisseurship which he applied to the idea of evaluation. (Many authorities have replaced the term "evaluation" by the term "assessment." I have retained the term "evaluation" in all my work). It simply means, not with-standing Eisner, a process for determining whether or not we have achieved our objectives. We can't help wanting to achieve something and in this script that is an objective.

Connoisseurship implies knowledge and skill that has been built up over time. It is an expertise, or as today's jargon would have it, a "competency." It is a skill that can be learned, and with beginning teachers it is one way they can begin to acquire the tacit knowledge that is so important in teaching. It is a way of reflecting on and bringing a critical eye to one's practice, that is, educational criticism. Eisner wrote: "As one learns to look at educational phenomena, as one ceases using stock responses to educational situations and develops habits of perceptual exploration, the ability to experience qualities and their relationships increases. This phenomenon occurs in virtually every arena in which connoisseurship has developed. The orchid grower learns to look at orchids in a way that expands his or her perception of their qualities. The makers of cabinets pay special attention to finish, types of wood and grains, to forms of joining, to the treatment of edges. The football fan learns how to look at plays" (set pieces in soccer), "defense patterns and game strategies. Once one develops a perceptual foothold in an arena of activity—orchid growing, cabinet making, or football watching—the skills used in that arena, one does not need the continual expertise of the critic to negotiate new works, or games or situations. One generalizes the skills developed earlier and expands them through further application" [8].

To develop this skill of connoisseurship I suggested to my trainee graduate teachers that at the end of the day they should reflect on what had happened in a class, by trying to visualize that class as an impressionist painting. I hoped it would help them understand (perceive) what had happened that was educationally significant. Exhibit 2.1 shows two reports from student teachers on what happened in their classes when their students were taught a problem solving heuristic. It might be argued that teacher (b) shows more insight than teacher (a) but this is not to say, that with further experience teacher (a) would not show an increase in insight, particularly if he/she had had sight of examples considered to have met the criteria.

The idea is to see the classroom in a different light. The skill of connoisseurship can only be developed with practice and conversation. To further develop the skill I asked my graduate trainees to provide me with an overall evaluation a week or so after the lesson had been conducted, and after they had analysed the results of the test they had designed to evaluate the strategy used. In the cases shown in Exhibit 2.2 they had been asked to evaluate a reported research on the effect of examples on teaching a concept (see Journey 10). To enable them to complete the final evaluation, I had provided them with a chapter from a book by Howard on concept learning in order for them to make a judgement based on theory and the evaluation practice.

Eisner's view that skill in educational criticism requires an adequate theoretical base was met by the provision of Howard's book. Eisner clarifies what he means by this when he relates it to reflective thinking which he regards as the base for curriculum thinking. He calls the reflective moments that a teacher has "preactive teaching," a term coined by P. W. Jackson. Such moments occur, Eisner writes, "prior to actual teaching; planning at home, reflecting on what has occurred during a particular class session, and discussing in groups ways to organize the program. Theory here sophisticates personal reflection and group deliberation. In so far as a theory suggests consequences flowing from particular circumstances, it enables those who understand the theory to take these circumstances into account when planning."

"In all of this, theory is not to be regarded as prescriptive but as suggestive. It is a framework, a tool, a means through which the world can be construed. Any theory is but part of the total picture… In one sense all teachers operate with theory, if we take theory to mean a general set of ideas through which we make sense of the world" [8].

"All teachers whether they are aware of it or not use theories in their work. Tacit beliefs about the nature of human intelligence, about factors that motivate children, and about the conditions that foster learning in classrooms. These ideas not only influence their actions, they also influence what they attend to in the classroom, that is, the concepts that are salient in theories concerning pedagogical matters also tend to guide perception. Thus, theory inevitably operates in the conduct of teaching as it does in other aspects of educational deliberation. The major need is to be able to view situations from the varied perspectives that different theories provide, and thus, to be in a position to avoid the limited vision of a single view" [9].

a. "As is normal in these research classes students are issued with the relevant handouts, and began beavering away. The first handout issued related to their emotional and motivational states (questionnaire), the second to their decision making models, the third was a repeat of the Payne experiment, while the fourth familiarized them with the daily decisions of science, the fifth tested the application of problem solving skills. The class was divided in half, one group were given the problem solving sheet (questionnaire 4) and were not shown how to subsequently solve the problem. A second group were tested and subsequently shown the correct way of solving the problem, and then both groups were retested. Discounting memorization this should (by comparative analysis of the scores of group A and B) illustrate if problem solving skills can be taught. Distribution of questionnaires takes up a large proportion of the student teacher's time in such a class and efficient organization is essential. The sequence of using the questionnaires together with a specified place for each completed set must be designated beforehand by the researcher. Sloppy organization makes for a badly run class and much time can be lost as a result. Students enjoyed playing with the information search 'card' sheet which was passed around the class and as students completed the questionnaires. The typing of the questionnaire by the school secretary was of immense benefit, as students were not strained into deciphering my handwriting as is usual, in these exercises where there is usually a large volume of hand written material. Students were patient in filling out the forms and in listening to the initial 'talk' given about Kolb, the importance of educational research and decision making skills to examination performance. They co-operated but class noise levels were abnormally high. This exercise extended across 4 classes and what has been included here is a general impression of all classes." (See Journey 12)

b. "The first problem set to the class was a chemistry problem and approximately half the class got it correct. I then asked the class to offer their views on what problems they might encounter in their lives. This led to a discussion on the various types of problems that first years (13 years of age) might typically encounter including exam questions, and to how to do away with teachers they didn't like."

"I then explained what a heuristic was, and outlined the steps in the heuristic class to them. We went through the steps in the heuristic and 'solved' the problem already given to them. One of the pupil's (Jordan) thought that four and five were the same, or at least there was not much of a difference between these two steps. Some of the others in the class said that they already used these steps and it was just stating the obvious. They were then set a second problem which they solved by themselves in class, using the heuristic. However, I am not totally convinced that they actually understood the heuristic to the problem. I think that maybe some of the pupils may have grasped the idea of a 'plan of action' in order to solve a problem, but the rest have not realized the significance of the heuristic and have not used it to help them solve the problem."

"Although the pupils seemed to enjoy the discussion of problems in general (which were not subject related) they did not appear willing (or maybe able) to apply the heuristic. I don't think this lesson was the most successful lesson I have ever taught."

Exhibit 2.1: Two reports from student teachers on what happened in their classes when their students were taught a problem solving heuristic.

(a) "In conclusion, I wish to say that the whole exercise was a very interesting process in which I learned something new about educational theory, my subject, my students, and myself. I feel that this single small experiment was an imperfect attempt to assess the theory on concept teaching, and with the benefit of hindsight I could probably design the lesson plan (and perhaps the test) so that it better assisted the attainment of behavioral objectives. The theory seems to be correct-perhaps useful is a better word—but I would hope to conduct experiments in the future using controls to arrive at a stronger verification."

"The second reading (Howard) appeared to offer little to alter the fundamental theory on the use of examples and non-examples and rather offered refinements on its use, in addition to some other techniques (use of metaphors and concept maps) to embellish it. From this I infer that the idea of using examples and non-examples has held its own over the years, and it is in my own teaching practice that I will have to investigate the value of the theory further. This will certainly require a restructuring of my approach to lesson planning as the methods that I have been using hitherto have been based more on intuitive feel than hard facts and experimental evidence."

"The most exciting prospect is that the classroom can be approached as experimental laboratory in which to apply, test, and evaluate ideas on how to improve students learning. The challenge to me as a teacher is to become active in being experimental and open to changing my preconceived or un-thought through attitudes on how to do things."

(b) "After reading Howard I would not have been afraid to use metaphors in the lesson. I was unsure of beginning the class with connecting percentages to decimals and fractions, but like everything else in mathematics everything is interconnected and cannot be understood until the lower steps have been mastered. At the end of the class they realized that percentages, fractions and decimals represented a part of a value and could tell the difference between them. I am familiar with students taking the incorrect aspects of the metaphor in learning a concept and I am conscious only to use metaphors with great care only after discriminating between the analogy and the new concept. I like the idea of concept mapping as this would allow me to work out at what stage the students are at and it could help me start at their level instead of having to make assumptions all of the time. If they didn't understand what I was presenting I would have a logical plan to refer back to. Likewise if they already knew what I was doing."

Exhibit 2.2: Extracts from the evaluations of lessons given by graduate student teachers in (a) science and (b) business mathematics in which they were asked to test the validity of specified research on the teaching of concepts using examples (see Journey 10). After they had completed their study they were given a chapter from Howard, R. W. (1987) *Concepts and Schemata. An Introduction*, London Cassell, and asked to take it account when writing their final evaluation.

If student is substituted for child(ren) in the above it will be seen to apply equally to higher education. It seems to me to be akin to Newman's "philosophical habit of mind" which is continually developing the skill of criticism, and is surely what Elliott means by practical reflection [10]. Newman wrote, "[…] a philosophical cast of thought, or a comprehensive mind, or wisdom in conduct of policy, implies a connected view of the old (*the teachers prior understanding*) with the new" (*the result of what happened in the class*); "an insight into the bearing and influence of each part" (*the students and the teacher*) "on every other; without which there is no whole and could be no center. It is the knowledge, not only of things" (*students and teacher*), "but of their mutual relations. It is organized and therefore living knowledge" [11].

One reason why trainee graduate teachers want their training programmes to concentrate on giving them tips for teaching rather than on theory, is that they come with years of experience of having been taught in one way or another. During that time they are forced to make judgements about what constitutes good teaching, and what does not, on the basis of that teaching. They acquire their own theories of effective teaching, and some of the theories that teacher trainers might propose may bring about cognitive dissonance [12].

One criticism of the idea of connoisseurship is that connoisseurship is about taste, and critics often disagree about taste. Therefore, evaluations should be based on empirically based knowledge [20]. But assessors may well disagree about a particular teacher's performance. A check list such as those that are often used may not grasp the teacher's performance as a whole. The key is the success or otherwise of the class in achieving its objectives, and the teacher's willingness to change, if change is perceived to be necessary.

There is also the problem that a particular theory may not be easily transferable to a particular class. Therefore, the function of the practical component of teacher education should be to enable beginning teachers develop their own theories of teaching and tacit knowledge through guided practice in which they evaluate a range of instructional strategies and theories. This can only be achieved by some kind of "research" which enables teachers to "discover that the classroom is, or should be, a challenging research laboratory, with questions to be pursued, data to be collected, analyses to be made, and improvements to be tried and evaluated." In that way the status of university teaching should also be raised with the development of a scholarship of teaching, so thought K. Patricia Cross [21]. James Trevelyan, a distinguished engineer goes one step further and argues that "it is helpful for many engineering faculty to understand that teaching expertise can help their research as well as classroom teaching" [22]. I will explain in Journey 3 how I asked my graduate student teachers to evaluate a range of theories and strategies of teaching, and so reflect on and develop their own theories.

NOTES AND REFERENCES

[1] Rokeach, M. (1960). *The Open and Closed Mind*. New York, Basic Books. 11

[2] Heywood, J. (a). (1989). *Learning, Adaptability and Change. The Challenge for Education and Industry.* London, Paul Chapman/Sage. (b). (2017). *The Human Side of Engineering.* Morgan & Claypool. Store.morganclaypool.com 11

[3] Elliott, J. (1976). Preparing for classroom accountability. *Education for Teaching*, 100, pp. 49–71. 11

[4] Heywood, J. and H. Montagu Pollock (1977). *Science for Arts Students: A Case Study in Curriculum Development.* Guildford, Society for Research into Higher Education. 12

[5] The original approach to perception is outlined in Ch. 4 of Heywood, J. (1982). *Pitfalls and Planning in Student Teaching.* London, Kogan Page. It was updated for continuing professional development courses in instructional leadership, and management in (a) *Instructional and Curriculum Leadership. Toward Inquiry Oriented Schools*, (2008). Dublin, National Association of Principals and Deputies. (b) *Managing and Leading Schools as Learning Organizations. Adaptability and Change*, (2009). Dublin, National Association of Principals and Deputies. Revised again for engineering students in *The Human Side of Engineering.* St. Rafael, CA, (2016). Morgan Claypool. 13

[6] Abercrombie, M. L. J. (1960). *The Anatomy of Judgement. An Investigation into Processes of Perception and Reasoning*, London (1989 reprint), Free Association Books. 13

[7] Eisner, E. (1979). *The Educational Imagination: On the Design and Evaluation of School Programs.* London, Collier Macmillan. 13

[8] *ibid* 13, 14

[9] *ibid* 14

[10] Newman, J. H. (1923 impression). *The Idea of a University: Defined and Illustrated.* London, Longmans Green. Discourse 5. 17

[11] Newman, J. H. (1843). Fifteen sermons preached before the University of Oxford. London, Rivington (1890 ed.), Sermon 14, p. 287. Also on pages 291 and 292. 17

"Philosophy, then is reason exercised upon knowledge; or the knowledge not merely of things in general, but of things in their relations to one another. It is the power of referring everything to its true place in the universal system -of understanding the various aspects of each of its parts, -of comprehending the exact value of each, -of tracing each backwards to its beginning and forward to its end, -of anticipating the separate tendencies of each, and their respective checks or counteractions; and thus of accounting for anomalies, answering objections, supplying deficiencies, making allowance for errors, and meeting emergencies."

[12] "An important characteristic of memory and perception is the tendency to remember our successes and forget our failures [13]. At the same time we tend also to be very consistent in our attitudes and opinions [14]. Apart from the fact that this makes it more difficult to adapt, we try to adapt, by accommodating new perceptions that possess values within our value maps. We tend to use sets that have served us well in the past. The same is true of problem solving: we tend to use the same heuristic whatever the problem [15]. Bruner has called this *persistence forecasting* and it can in a new situation prevent us from using more efficient strategies. We tend to believe in the advantages of what we already possess. Dissonance or downshifting arises when we have to accommodate a new value system with which we have no empathy." When there is a conflict between the values of the instructor and the student learning may be impeded. 17

"For example when a subject that we have to learn is cognitively complex and where values are involved, such as in political studies, a student may be in disagreement with the views held by the teacher. In these circumstances there may be considerable resistance to learning, and apparently this is particularly likely to be the case if the students are only mildly critical of the teacher's standpoint. Such students may become alienated from the political and economic system. However, as Marshall [16] shows, a teacher can cause learning through his teaching style, even if his or her rating with the students deteriorates during the course." (Which is one of the problems of using student ratings to evaluate the effectiveness of a teacher).

"Since we impose meaning on the objects of knowledge it should come as no surprise to find that a [student] can deliberately impose misunderstanding in order achieve consistency between the message and his feelings. If there is consistency a student can change his/her attitude to a teacher from like to dislike if the teacher's messages appear to be untenable. This can happen in university when first-year [engineering] students have to cope with certain propositions in the social and behavioural sciences: anything that is contrary to the student's views can create such dissonance." [...] [17].

"Challenges to values may be perceived as threatening. More generally in situations perceived to be threatening, we narrow our perceptual field and return to the safety of our beliefs [18]. Behaviour in which we revert to tried and trusted ways can affect the higher order cognitive functions and thus the ability to solve new problems. Downshifting of this kind it is argued is one of the reasons why students fail to apply the higher levels of the Bloom Taxonomy" [19].

"Cognitive dissonance theory accounts for the behaviour of institutions and politicians. For example, politicians become so committed to the values expressed in their slogans that they become unable to entertain reasoned arguments against their points of view even from some of their own supporters! And of course it is necessary for the dynamic

of political parties that their workers (supporters) should not deviate from their beliefs"
[...]

The above paragraphs are from pages 66 and 67 of Heywood, J. (2008). *Instructional and Curriculum Leadership. Towards Inquiry Oriented Schools*. Dublin, National Association of Principals and Deputies.

[13] Bruner, J., Goodnow, J. J., and G. A. Austin (1956). *A Study of Thinking*. New York. Wiley. 19

[14] Festinger, L. (1959). *A Theory of Cognitive Dissonance*. Stanford, CA. Stanford University Press. 19

[15] Luchins, A. S. (1942). Mechanisation in problem solving: The effect of "einstellung." *Psychological Monographs*. No 248. See also McDonald, F. J. (1968). *Educational Psychology*. Belmont, CA. Wadsworth. 19

[16] Marshall, S. (1980). Cognitive affective dissonance in the classroom. Teaching political science in the classroom. *Teaching Political Science*, 8(11), pp. 111–117. 19

[17] French, W. L., Kast, F. E., and J. E. Rosenzweig (1985). *Understanding Human Behaviour in Organizations*. New York, Harper and Row. See also Thouless, R. H. (1974). *Straight and Crooked Thinking*. London, Pan Books. 19

[18] Combs, A. W. and D. Snygg (1949). *Individual Behaviour. A Perceptual Approach to Behaviour*. New York, Harper and Row. 19

[19] Caine, R. N. and G. Caine (1991). *Teaching and the Human Brain*. Alexandria VA, Association for Supervision and Curriculum Development. 19

[20] Schubert, W. H. (1997). *Curriculum. Perspective, Paradigm, and Possibility*. Upper Saddle River, NJ, Prentice-Hall, p. 276. 17

[21] Cross, K. P. (1986). A proposal to improve teaching or "what taking teaching seriously should mean." *AAHE Bulletin*, p. 14. 17

[22] Trevelyan, J. (2010). Engineering students need to learn to teach. *ASEE/IEEE Proceedings Frontiers in Education Conference*, F3H-1 to 6. 17

2.5 APPENDIX

Example of an action research in a primary (elementary) school initiated by a teacher with the support of the Principal of his school.

This extract is from Heywood, J. (2008). *Instructional and Curriculum Leadership. Toward Inquiry Oriented Schools*. Dublin, National Association for Principals and Deputies, pages 45 ff.

It is based on Prior, P. (1985). *Teacher Self-Evaluation using Classroom Action Research*. A Case Study. M.Ed. Dissertation. University of Dublin, Dublin.

Pat Prior set out to establish the validity of Elliott's model of action research in an Irish primary school. He did this with the support of his Principal and his colleagues on the staff. As defined by Elliott action research is "the study of a social situation with a view to improving the quality of action within it." Elliott, J. (1991). *Action Research for Educational Change*. Milton Keynes/Philadelphia, Open University Press.

The start of any research is to clarify the problem and this is often by no means easy. In the first place Prior used, as did Elliott the Nominal Group Technique. This is an extension of brainstorming. It has six stages: (1) Question setting. (2) Reflection. (3) Pooling. (4) Clarification. (5) Evaluation and (6) Review. It is quite an extensive procedure and Prior found that in order to clarify the problem he required more than one session.

One effect is that it requires teachers to publicly declare problems they have and in so doing they have to try and separate self-esteem from classroom activity. Elliott found that teachers find this difficult to do. Exercises like this also demonstrate that the problems that teachers face in isolation are likely to be the case in other classrooms. Prior's study found that his school was no exception.

Prior circulated a summary of the first meeting. Before the next meeting he also circulated a document that set down the aims and guidelines for future meetings. At this stage the purpose was seen to be "to make a largely academic curriculum more meaningful for the development of the whole student."

At this stage the investigator also recorded the inability of the project to alter the academic emphasis of the curriculum. The principal said that nothing could be done about it. "Efforts should be made to make the academic curriculum more attractive to academically less able students." Nevertheless, Prior reported that the variety of issues that had been raised offered scope for classroom research. The teachers had to learn to focus. To get this focusing Prior decided to intervene during the third meeting with the teachers through means of a statement and a question. Comparison between the records of the earlier meeting and the meeting suggested to Prior that at last the teachers had begun to focus.

The next stage in the Elliott model is reconnaissance. It has two components-description and explanation of the problem. The problem now became that (1) some pupils were not fully occupied in the class, and (2) how could this situation be improved? The first would be for research the second would be for discussion meetings. Prior now asked his colleagues to observe and record what happened in their classrooms. The questions to guide observation are shown in the table. The written results ranged from the diary type listing of events to reflection. The general conclusion was that teachers lessons aimed at the average group and that those above or below this range were neglected.

At this stage Prior asked the teachers to shadow study a single pupil and to invite observers into their rooms (triangulation). Tape recorders were also provided. The principal assisted by supervising the class of a teacher who was observing. But at this stage Prior records that a crisis occurred (in so far as the investigation was concerned) because what was happening was not leading to change. He, therefore, decided to continue at a lower level and to concentrate on change in classrooms rather than at school/staff level.

Ten hypotheses had emerged as a result of classroom observation. It was accepted that there was a serious problem that centered on teacher's problems in dealing with students of all ability levels. It was proposed that there should be an in-service day to discuss this problem, but for a number of reasons including the failure of the university (this writer) to come up with a facilitator it never ran. There were now only 5 weeks left. So Prior decided to ask teachers to become researchers in their own classrooms and to devise, implement and evaluate a lesson that took into account the earlier findings and the hypotheses they generated. The results were of some interest. One teacher whose study is fully recorded was very successful but found the exercise exhausting.

Teachers were also asked to submit any aspect of learning that interested them; an interview with the principal was also conducted.

Prior points out that while many teachers said that change was not possible because of class size, one teacher had actually achieved such change.

This is by no means all but that is not for this text. While the project met with many vicissitudes there is no way that it can be regarded as a failure. It may not have achieved its goals but like any action research it achieved many things en-route. First, it showed that many teachers find it difficult to reflect and have to be helped if they want to achieve a new level of thinking. Second, the whole process is lengthy. Prior felt that if he were to do it again he would shorten the process of finding the problem. Perhaps, however, teachers should go through that lengthy process. Third, at the time he wrote he did not see a direct connection with the whole school plan. When he undertook the project whole school planning was in its infancy. Similarly, research on TQM (Total Quality Management) in educational institutions was only beginning to appear in books and journals. It seems clear from this research that a school that concentrates on a project like this over a year is likely to achieve much more than the engagement of small groups in different projects. Here the whole school was involved in the problem of teaching mixed ability groups, a problem that is still with many teachers. It is in this sense that we begin to understand the concept of teamwork in schools. Fourth, the project indicates, […] that there is much more to the design of instruction than is currently thought to be the case. Fifth, the conduct of the project provided a *chocs des opinions* and began to get the teachers thinking outside of their normal frames of reference. Instructional leaders will find the maintenance of such attitudes difficult unless agreed changes are built into the curriculum. Sixth, although teachers believed they were constrained by a received curriculum and large class sizes they nevertheless were able to undertake developments within these constraints. In all, during this period the school was of the kind described by Cohn and Kottkamp, that is inquiry oriented with teachers acting as extended professionals.

Aim
To gather evidence from classrooms which is relevant to the following problem: there are some pupils who are not fully occupied in class. How can this situation be improved?

Guidelines
Teacher asked to observe and describe the fact of the situation, using the following suggestions:

Which pupils are not adequately occupied in class?

What are such pupils doing when they should be working?

Do pupils so behave during a particular type of class or teaching or at certain times of the day?

Any other instance and circumstances of time wasting noticed.

Exhibit 2.3: This extract is from pages 313–317 of Heywood, J. (2009) *Managing and Leading Schools as Learning Organizations. Adaptability and Change*. Dublin. Original Writing for the National Association of Principals and Deputies. The information on Prior's course is taken from Prior, P. (1985). *Teacher Self-Evaluation using Classroom Action Research*. Dublin, M.Ed. Thesis. School of Education, University of Dublin.

JOURNEY 3

Toward a Scholarship of Teaching. Teaching as Research

3.1 INTRODUCTION

In the middle 1950's the only mechanism for training teachers in technical colleges in England was day release to a Technical Teacher Training College of which there were four scattered across the Country. Since technical teachers were not required to be trained those who obtained day release for training were fortunate. There were, however, the little known qualifications of the College of Preceptors that could be taken by self-study and examination. The College was the first organization created for teacher training in the UK. It was incorporated by Royal Charter in 1849. By the mid-twentieth century it offered an associate diploma, a degree level licentiateship, and higher degree level fellowship that were intended for practising teachers in the school sector. The examination for the licentiateship was in two parts, subsequently extended to three: the first part might be described as the principles of education, the study of philosophy, psychology, history of education and administration. The second part related specifically to teaching and the subject that was being taught by the candidate; in addition to a written examination it required a 10,000 word dissertation related to the candidate's experience of teaching during the preceding year; the third part consisting of three papers related to knowledge of the subject taught by the teacher. Strangely enough, given the focus on secondary education, part 2 could also be taken in "technical education" for which there was a specific subject examination to which the dissertation had to be related. In this part of the examination the College was encouraging teachers to become researchers into their own instruction, and in so doing to extend their "tacit" knowledge. This is how I qualified as a technical college teacher, and became interested in research in technical education. I extended that interest by taking the fellowship.

The College recognized, as we are now beginning to recognize, that lectures were unnecessary if all that they do is to repeat what is already in textbooks. Any person sufficiently motivated could read the textbooks and learn to answer the examination questions. Correspondence colleges were quick to offer programmes to support candidates who valued tutorial support. The college also recognized the need to value practice, hence the dissertation in part two of the Licentiate. It was this experience as a technical college teacher, some 25 years earlier, together with my failure in Ireland to get the graduate trainee post-primary teachers in my course in the Applied Psychology of Instruction to relate theory to practice, and *vice-versa* that caused me to change the course into a series of action based inquiries. They were now required to evaluate certain instructional strategies and theories in their classrooms [1].

As it stood, the programme required the graduate-student teachers to prepare a huge number of lesson plans. These generally contained a script for a lesson, often a modification from a textbook and little else. I had no means of knowing whether the students used them in this way or not, and came to the conclusion that they were rather a waste of time.

I also wanted to improve the approach to lesson planning because children show they recognize when the organization of teaching and learning impedes the attainment of goals; that is, if they are asked, which more often than not, they are not. One has only to look at children at play to see that many of them bring structure and organization to what they are doing. It helps their learning. One of the judgements they make about teachers is the degree to which they organize what is to be taught, and how it is to be learned. They do not want their teachers to be chaotic.

3.2   THE SCHOLARSHIP OF TEACHING

Students in higher education are no different. Lessons and lectures have to have a degree of planning if they are to be perceived as successful. Evaluation, if done properly, should yield insights that develop the teachers tacit or working knowledge. Evaluation and reflection turn teaching into a continuing activity of inquiry and, sometimes research. That is the "scholarship of teaching."

There were two premises behind this approach. First, the teachers were forced to take the strategies and theories of instruction that would normally have been given in lectures, try them out in the teaching situation, and determine their value for them. Second, for the trainee teachers to develop critical skill in observing the "happenings" in their classrooms, in order to prepare them to systematically investigate those they viewed as critical for the development of their teaching. In this way the trainees were forced to challenge their own pre-conceived theories which might be strengthened or changed as a result of these experiences. In so doing they would acquire a system of tacit knowledge.

The process of each activity is outlined in Exhibit 3.1. It does not include the tutor's activity in grading, or giving feedback on the assignment which is integral to the process. When the course began the five activities to be investigated during the year were (1) The teaching of concepts using examples. (2) The use of imagery in instruction. (3) The value of knowledge of student learning styles to the tutor. (4) The relative merits of expository teaching when compared with one or another form of discovery (inquiry) learning. (5) The merits of teaching a problem solving/decision making heuristic in problem solving and learning [2].

The process might be described as pseudo-scientific experiment. The overall activity is no different to any other design activity as Exhibit 3.2 shows. The teacher had to design a lesson to

1. Academic Course: Introduction to activity (2 – 4 hours)

2. Student Preparation

   a. Read the literature on the designated topic (provided)

   b. Select a small topic from the literature for investigation (this may be to replicate one of the studies reported in the literature).

   c. Design a lesson to test the hypothesis shown in (b); (this to include the entering characteristics of the pupils, a statement of aims and objectives, the instructional procedures showing how they will test the hypothesis, etc.)

   d. Design a pupil test of knowledge and skill which is directly related to the objectives of the lesson.

3. Academic Course: (only if students require a seminar) to iron out difficulties (2 hours).

4. Student Implementation

   a. Implement Class as designed.

   b. Immediate Evaluation. (Evaluation 1)

      i. What happened in the class?
      ii. What happened to me?
      iii. What have I learned about myself?
      iv. What have I learned about my pupils?

5. One Week (or so) Later.

   a. Test Students- comment (Evaluation 2).

   b. Substantive evaluation (evaluation 3)

      i. How does what I have done relate to the theory which I set out to evaluate?
      ii. How, if at all will this influence my teaching in the future?

6. Submit report at the required time.

Exhibit 3.1: The process of the "lesson plan" activity (Heywood, J. (2007). *Instructional and Curriculum Leadership. Toward Inquiry Oriented Schools*. Dublin, Original Writing for the National Association of Principals and Deputies—Chapter 5).

test the hypothesis that they had taken or deduced from the literature. Then they had to design a test that is related to the content of the course as normally taught, and at the same time, they had also to evaluate the method (theory) of instruction used: A considerable task. The test had to be subjected to descriptive statistical analysis, and a substantial written evaluation had to be made.

I would argue, that these studies meet Elliott's requirement for action research because they "study the social situation (classroom) with a view to improving the quality of action within

Simplified Model of the Engineering Design Process	Lesson Planning Process		
Vague statement of what is wanted		Knowledge	
Problem formulation	Problem formulation	Cognitive skills	
Broad view of problem		Affective skills	
Problem analysis		Assumptions	
Details of problem	Alternative solutions		
Search	Statement of Aims and Objectives		
Many partial solutions mainly in concept form			
Decision	Construction of Lesson Plan		
Preferred solution	Implementation of Lesson	Feedback	
Specification			
Details of proposed solution	First evaluation		
Model	Test		
Evaluation	Analysis of Data		
Manufacture	Final Evaluation		

Exhibit 3.2: The lesson planning process contrasted with a simplified model of the engineering design process due to E. V. Krick (1969). *An Introduction to Engineering and Engineering Design*, 2nd ed., New York, Wiley. Page 156.

it" [3]. Because it was action research it was not expected that the student teachers should stick strictly to the plan irrespective of difficulties experienced in the classroom. What mattered was that satisfactory reasons were given for the change in plan. The progress of teaching is often non-linear. It will be seen that Section 2 requires the student to do the reading that would normally have been given in support of a lecture on a particular theory, or strategy of instruction. Sections 2(a) and (b) required an extensive essay response.

Nevertheless, it took me several years before I abandoned giving an introductory lecture and replaced it by a dialogue session among the 100 or so students taking the course. It also took me several years before I abandoned the traditional 2/3 hour terminal examination that had been set at the end of the year in favor of a written examination in which one prior notice question given out at the beginning of the course had to be answered. It had the intention of obtaining the second goal of the course (see Exhibit 3.3).

But, Section 2 also requires the student to take an entirely different approach to lesson planning. They are asked not only to prepare a lesson to meet the requirements of that part of the school curriculum with which they are dealing, but design it in such a way that it would

Write an essay on the following:

In your lesson plans you undertook investigations which replicated previously published research on learning in the classrooms. However, one of the goals of the course is that you should be able to design investigations which help you better to understand the significant events which you experience in the classroom. Describe any significant event which you have experienced that still requires explanation and suggest procedures for its investigation. Give a detailed example of such research. (You may not use material from your assessed lesson plans in this answer).

Exhibit 3.3: The seventh exercise which had to be completed either as an essay at the end of the term, or as a one hour terminal paper. Previously it had been set as an examination only. The question was given to the students at the beginning of the academic year (October) and examined in June. (Examples will be found in Heywood, J. (1992). Student teachers as Researchers of Instruction in the Classroom in J. H. C. Vonk and H. J. van Helden (Eds.). *New Prospects for Teacher Education in Europe*. Amsterdam, Free University. Association for Teacher Education in Europe).

test the theory or practice they were required to evaluate. It was a major task for a beginning educator. It was one that I later found was equally difficult for experienced teachers in continuing professional development programmes. Add to that the demand that they should design a test that would not only test the routine matters of the curriculum, but the theory or practice that they wished to evaluate. It was expected that this test would be administered a week or so after the lesson had been delivered.

Immediately after the lesson had been given the students were asked to evaluate what had happened (see also Journey 2). As Exhibit 3.1 shows they were asked to say "what happened in the class?" "What happened to me?" "What have I learned about myself?" And, "What have I learned about my pupils?" I certainly hoped that they would see this as an opportunity to practice educational connoisseurship, and some students tried to do just that.

You may take the view that the example shown in Exhibit 3.4 demonstrates such connoisseurship. It shows that sometimes teachers misjudge the effects of a class. In this class the teacher had decided to design a lesson in English (poetry) that catered for the needs of the four different learning styles described by David Kolb [4, see Journey 12]. She chose it so that it could be linked to the problem of bullying which had been raised with her by this class of 12 year old boys. A second point is that students sense what is happening to teachers; perhaps more readily than we would care to believe.

This activity had two purposes, one that I understood at the time, and one which I learned, The purpose that I understood at the time was the objective of getting these student graduate educators to stop worrying about themselves, and begin to try and understand what was happening to their students, and why. It is only recently that I have begun to realize that this activity contributed to the development of their tacit knowledge, and that it is this tacit knowledge that forms the basis of their technical knowledge and theories of learning.

Evaluation 1

The one thing that struck me even while I was teaching the lesson was the mute reaction of the class to all parts of this lesson. They seemed to exhibit little enthusiasm for anything. They are a lively bunch, and I had thought that this poem and topic would be perfect for them. They have just come out of the primary (elementary) school scene so the experience would be relatively recent for them and they had previously brought up the topic of bullying voluntarily. The presence of a supervisor from TCD could have been a contributing factor but this had coincided before with a lesson plan session (lesson plan 1), and a glance at their performances in that class proves that this would not dampen their spirits. My supervisor, who was only present for the first class, even commented on their very quietness, or had they had a class before mine which repressed their behavior, which incidentally, did not dramatically change after she left. Perhaps it was just an off day for them. . My own performance on the day could also have been a contributing factor. On that particular day I was quite sick so perhaps the students sensed that I wasn't fully with them.

But however disappointed I was with the response of the class to the lesson, their results indicate that it was their best performance so far (continues....)

Exhibit 3.4: Example of a first evaluation by a student teacher.

Although the groups were small, anything from a 12 to 30 pupils, it is nevertheless useful for trainees to do some basic statistics such as the mean and standard deviation of the test results, if only to make them think about what is happening in the class. It is a second evaluation, but of a very different kind. Since, it is undertaken a week or so after the lesson it gives the student an opportunity to re-consider what happened, and what they have learned that will inform their future teaching. This activity was always called the third evaluation, but some students used the term "reflection," and for many it went beyond evaluation to reflection.

Sometimes the second evaluation (Exhibit 3.5) ran into the third (Exhibit 3.6). In this case, the teacher had designed two classes to evaluate the relative merits of expository and discovery teaching. It was a science class on materials. He had found that the guided discovery class were highly motivated, but that it did not seem to be, "enough to promote understanding as demonstrated by the results." He also found that brighter and middle range students responded better to the inquiry approach whereas the weaker students seemed to benefit more from the expository approach; "to a large extent I feel that the teacher has the ability to motivate and create interest in a class with even the most tedious topics." This extract from his report is intended to illustrate that in doing this kind of action research the teacher who is both observer and participant should be aware of the assumptions that have to be made, and to understand the limitations on the illumination that the exercise might provide.

As indicated in Journey 2 in order to help students develop skills for the final evaluation, with some exercises I gave the students a reference, and asked them to take it into account when considering what they would do differently if they had the opportunity to repeat the exercise again. The first exercise always asked the students to replicate one of the experiments on the role of examples and non-examples in the teaching of concepts. Exhibit 3.7 is part of an evaluation of the teaching of two classes the same concept of animal cells, but using different approaches to the sequencing of the examples. As indicated in Journey 2 the students were asked to read a text by R. W. Howard on concept learning [5]. Exhibit 3.7 shows how a teacher's position changed after reading Howard. Of some interest is the recognition that the classroom was a laboratory for research. In my classification this evaluation belongs to the class of technical evaluation and not reflection.

3.3   TEACHING AND DESIGN

These activities are little different to design and make projects. Describing the key skill of technical coordination in project management Trevelyan wrote, "First the engineer describes what needs to be done and when, and negotiates a mutually agreeable arrangement with other people who will be contributing their skills and expertise. Next, while the work is being performed, the engineer keeps in contact with the people doing the work to review progress and spot misunderstandings or differences of interpretation. The engineer will also join in discussion of unexpected issues that arise, and may need to compromise on original requirements. Third, when the work has been completed, the engineer will carefully review the results and check that no further work or rectification is needed."

While this may be seen as project management work, "technical coordination is an undocumented informal process that relies on personal influence, rather than lines of formal authority. This corresponds closely with pedagogy: first setting the task for students, secondly monitoring the students as they perform the task, offering help and guidance when needed, elaborating on the requirements of the exercise when the students misunderstand, and finally checking the student's work and assessing it against criteria that define relative levels of performance" [6]. It is within the process that the teacher negotiates his/her way through the lesson, making key decisions on whom and what to give time to, and the important decision as to whether to achieve the goals that were set in the lesson plan. Trevelyan's point is that much engineering requires social interaction, explanation and negotiation. Teaching, he argues involves all three, therefore engineering students should be given the opportunity to teach during their experience of college. "Teaching relies on accurate listening to understand the learner's needs, gaining the willing cooperation of the learner, planning expectations and assessment, presenting carefully planned explanations, and observations of human behavior and responses. These are the same social interaction skills that form the foundations of most professional skills we have observed engineers using" [7]. For this reason I believe that faculty have much to learn from beginning teachers but

[...]

In reaching this rather tentative conclusion I have had to make several assumptions which have to be seriously considered before these results could be considered valid.

1. That the questions in the exam tested the two methods. I do feel that this was a good exam and that it encouraged the students to explain what they understood by the topic. However, one is assuming firstly that a child can accurately put into words something that has been understood, and all pupils' language skills are equal. Also it is still hard to tell whether answers had been remembered but not understood. The students are very eager in general and have had several surprise tests to date. The combination of these two facts means that work is constantly covered at home by more eager students in the event of an exam. Hence, even in the questions designed to test understanding, pure recall may be being used. It is very difficult to tell.

2. It is also assumed that the students in class IY did not know about the test. Their science class is always one day after the IX class and hence it would not have taken much for the class to guess that they might have a test. Again, it is possible that some of the more eager students revised for this exam. Also some of the results actually increased from test one to test two in both IX and IY indicating, I feel more revision on the part of the class in the interim period which again means that long-term recall was not tested.

3. That there was no overlap in the two methods. This, for me at least, is a very grey area. How much guidance does one use in guided discovery? When does it stop being guidance and start being exposition? Until more exact definitions exist for these methods it will be very difficult to know exactly what one is testing for.

4. That the novelty factor, as previously discussed, is not what is driving the motivation rather than a heart-felt desire for knowledge and truth.

5. Finally, one is assuming that the statistics are valid. This itself is quite large assumption. In order to make any valid conclusion, setting aside the inherent problems in the method discussed already, one would have to repeat the exercise many times in order to get a significant sample. It would be wise to repeat it, using different topics with the same classes and then with different classes. It would have to be investigated to see whether this is age specific, gender-specific, race-specific and so on. To hazard a guess at this I would feel that perhaps boys would benefit more from guided discovery in science, based on experience of teaching boys I found them prone to throwing themselves into laboratory work and to wanting to do things themselves. It was nearly a sign of weakness to ask a teacher for help. I am not entirely sure that age would have an effect once the subject matter was suitable for that age group. (continues in exhibit 3.6)

Exhibit 3.5: Example of a second evaluation that ran into the third evaluation shown in Exhibit 3.6.

(continued from exhibit 3.5.)

As I result of this exercise I have, firstly, gotten to know my first year classes better which cannot have been a bad result. Secondly, it has made me look at the way in which I approach teaching classes and how perhaps one approach all the time is not ideal, even if just to tap into the novelty factor occasionally I feel that I would now tend to use discovery more than before and have found ever since that I am reluctant to tell classes what the outcome of an experiment should be. I prefer if they at least try to discover that for themselves.

Finally, it has made me realize how difficult a lot of Junior Certificate scientific concepts are in general and how frustrating this subject must be for those who would certainly appear, as a result of this exercise, to spend much time in the dark where science is concerned. Perhaps this is the most important piece of insight that I gained and it has certainly given me food for thought.

Exhibit 3.6: Example of a third evaluation which is a continuation of the second evaluation shown in Exhibit 3.5.

for all that, they should not expect students to undertake what they themselves are not prepared to undertake.

3.4   TEACHING AS RESEARCH–AN APPROACH TO SCHOLARSHIP

When the studies reported in this text were began in 1984 there were no texts that could be given to the students, moreover the approach was greatly influenced by my experience of engineering projects. This is no longer the case. Several texts have been written about teaching as research and classroom research on both sides of the Atlantic [3, 8, 9]. Two of the texts that have interested engineering educators have come from Patricia Cross and her colleagues Tom Angelo [10] and Mimi Steadman [11]. The first describes some fifty simple techniques for assessing what is going on the classroom. These are listed in the appendix. Many engineering educators use the "One minute Test." The second is an advocacy of qualitative research in classrooms with practical examples. There is no one perfect way in achieving what is probably better called "the scholarship of teaching."

In practice changing or developing the curriculum, irrespective of the level in the education system, is a design activity which is very similar to planning a lesson.

If I were to teach this concept again I would emphasize the non-examples more with weaker students. In order to do this I would use the idea of concept mapping as illustrated in the Howard reading. This is a process frequently used in science and would have proved extremely useful of the concept of the animal cell. A map of its place in the human system and a map of its place in the environment, including non-examples, such as rocks, within the environment, would possibly have given the class a much clearer picture to focus on.

In addition, I would have used the idea of highly typical examples (Mervis and Pani, 1980) to more effect, again for weaker students. Also, I feel that I did not make effective use of metaphors and analogies. I could have arranged the class into their own version of cells, with several grouping together to form a nucleus cell membrane, etc. Also they could have been given material to make 'their own cells,' e.g., a balloon filled with jelly and rubber ball inserted into the balloon, acting as a nucleus. Clearly, due to time constraints this would mean leaving out much of the practical work in the original lesson plan, such as the microscope work or overhead projector work. However, I would be very interested to try these new ideas and compare the results. The Howard article definitely opens up new opportunities in addition to those considered in de Cecco and Crawford (the text issued to the class for the exercise).

[…]

Hence, my behavioral objectives were met to a large extent with both classes but, as mentioned, the weaker students could well have benefited from a different approach.

Overall this was a very useful exercise in methods of teaching and has certainly opened my mind to new approaches. I will definitely incorporate the ideas into my classes from now on. Finally, this exercise has shown me how the classroom is a place for the teacher to learn and experiment also. I feel that I have a golden opportunity this year with two science classes running in tandem to investigate as many new approaches as possible.

Exhibit 3.7: Example of a final evaluation that was influenced by additional reading that provided at the time of the evaluation.

NOTES AND REFERENCES

[1] Heywood, J. (1992). Student teachers as researchers of instruction in the classroom in J. H. C. Vonk and H. J. van Helden (Eds.). *New Prospects for Teacher Education in Europe*. Amsterdam. Free University. Association for Teacher Education in Europe. 26

[2] Heywood, J. (2008). *Instructional and Curriculum Leadership. Toward Inquiry Oriented Schools*. Dublin, Original Writing for National Association of Principals and Deputies. Chapters 5–9 inclusive. 26

[3] Elliott, J. (1991). *Action Research for Educational Change*. Milton Keynes, Open University Press. 28, 33

[4] Kolb, D. A. (1984). *Experiential Learning*. Engelwood Cliffs, NJ, Prentice Hall. 29

[5] Howard, R. M. (1987). *Concepts and Schemata: An Introduction*. London, Cassel. 31

[6] Trevelyan, J. (2010). Engineering students need to learn to teach. *ASEE/IEEE Proceedings Frontiers in Education Conference*, F3H-1 to 6. 31

[7] *ibid* 31

[8] McKernan, J. (1960). *Curriculum Action Research. A Handbook of Methods and Resources for the Reflective Practitioner*, 2nd ed., London. Kogan Page. 33

[9] Phillips, D. K. and K. Carr (2006). *Becoming a Teacher through Action Research*. New York, Routledge. 33

[10] Angelo, T. A. and K. P. Cross (1993). *Classroom Assessment Techniques*. San Fransisco, Jossey-Bass. 33

[11] Cross, K. P. and M. H. Steadman (1996). *Classroom Research. Implementing the Scholarship of Teaching*. San Fransisco. Jossey-Bass. 33

3.5   APPENDIX

The 50 classroom assessment techniques described by T. Angelo and K. P. Cross in *Classroom Assessment Techniques*. San Fransisco, Jossey Bass, 1993.

Techniques for Assessing Course-related Knowledge and Skills
Assessing prior knowledge, recall and understanding

1. Background Knowledge Probe.

2. Focused Listing.

3. Misconception/preconception Check.

4. Empty Outlines.

5. Memory Matrix.

6. Minute Paper.

7. Muddiest Point.

Assessing skill in analysis and critical thinking

8 Categorizing Grid.

9 Defining Features Matrix.

10 Pro and Con grid.

11 Content, Form and Function Outlines.

12 Analytic Memos.

Assessing skill in synthesis and creative thinking

13 One-sentence Summary.

14 Word Journal.

15 Approximate Analogies.

16 Concept Maps.

17 Invented Dialogues.

18 Annotated Portfolios.

Assessing skill in problem solving

19 Problem Recognition Tasks.

20 What's the Principle?

21 Documented Problem Solutions.

22 Audio and Video Taped Protocols.

Assessing skill in application and performance

23 Directed Paraphrasing.

24 Applications Cards.

25 Student-generated Test Questions.

26 Human Tableau or Class Modeling.

27 Paper or Project Prospectus.

Techniques for Assessing Learner Attitudes, Values, and Self-awareness
Assessing students' awareness of their attitudes and values

28 Classroom Opinion Polls.

29 Double-entry Journals.

30 Profiles of Admirable Individuals.

31 Everyday Ethical Dilemmas.

32 Course-related Self-confidence Surveys.

Assessing students' self-awareness as learners

33 Focused Autobiographical Sketches.

34 Interest/Knowledge? Skills Checklists.

35 Goal Ranking and Matching.

36 Self-assessment of Ways of Learning.

Assessing course-related learning and study skills, strategies and behaviors

37 Productive Study-time Logs.

38 Punctuated Lectures.

39 Process Analysis.

40 Diagnostic Learning Logs.

Techniques for Assessing Learner Reactions to Instruction
Assessing learner reactions to teachers and teaching

41 Chain Notes.

42 Electronic Mail Feedback.

43 Teacher Designed Feedback Forms.

44 Group Instructional Feedback Technique.

45 Fundamental Assessment Quality Circles.

Assessing learner reactions to class activities, assignments and materials

46 Recall, Summarize, Question, Comment and Connect.

38 3. TOWARD A SCHOLARSHIP OF TEACHING. TEACHING AS RESEARCH

47 Group-work Evaluations.

48 Reading Rating Sheets.

49 Assignment Assessments.

50 Exam Evaluations.

JOURNEY 4

Objectives and Outcomes

4.1 THE SOCIAL EFFICIENCY IDEOLOGY

The social efficiency ideology requires that the curriculum serves utilitarian purposes, namely the creation of wealth. Institutions have to be run like businesses: therefore, the curriculum has to be seen to be providing measurable outcomes in the form of objectives now called outcomes. In this paradigm the teacher's role is to guide (manage, direct and supervise) the learner to achieve the outcomes (or terminal performances) required. Knowledge is defined behaviorally in terms of what a student "will be able to do," as a result of learning. It is the prevailing curriculum ideology in engineering education, as seen for example in the current ABET philosophy. The social efficiency ideology has its origins in the objectives movement.

4.2 THE OBJECTIVES MOVEMENT

The idea of stating educational objectives can be traced back to the end of the 19th Century. In its present day form it begins with the publication in 1949 by Ralph Tyler of a small book with the title *Basic Principles of Curriculum and Instruction* [1]. In it he set down four basic tasks for the educator. These are:

(i) The determination of the objectives which the course (class, lecture) should seek to obtain.

(ii) The selection of the learning experiences that will help to bring about the attainment of those objectives.

(iii) The organization of those learning experiences so as to provide continuity and sequence for the student, and help him/her integrate what might otherwise appear as isolated experiences.

(iv) The determination of the extent to which the objectives are being achieved.

It will be seen that assumptions are made about the content to be covered. The second task will depend in no small way on the beliefs that the educational unit responsible for the curriculum has about the delivery of that content, that is, the ideology it holds. The third task highlights two of the big problems faced by engineering educators. The first, is that of ensuring that the separate courses that make up the curriculum present a coherent picture of what engineering is all about [2]. The second, is the amount of time that should be devoted to each activity. There are continuing demands from engineering educators to increase the content at the expense of nothing [3].

Apart from the fact that these four tasks are not carried out in a linear fashion, they cannot be implemented without knowledge of the students entering characteristics, this being particularly important when diversity among students is being sought. The process may be expressed diagrammatically as in Exhibit 4.1. Today, especially in the U.S. this diagram would take into account programme objectives and show how they are aligned with learning objectives, or outcomes as commonly used today [4]. The remainder of this chapter documents a brief history of how the notion of objectives was developed for college and university examiners following Tyler's definition of an objective in 1949.

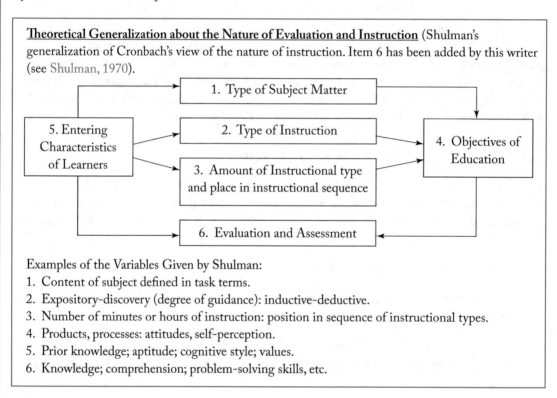

Theoretical Generalization about the Nature of Evaluation and Instruction (Shulman's generalization of Cronbach's view of the nature of instruction. Item 6 has been added by this writer (see Shulman, 1970).

Examples of the Variables Given by Shulman:
1. Content of subject defined in task terms.
2. Expository-discovery (degree of guidance): inductive-deductive.
3. Number of minutes or hours of instruction: position in sequence of instructional types.
4. Products, processes: attitudes, self-perception.
5. Prior knowledge; aptitude; cognitive style; values.
6. Knowledge; comprehension; problem-solving skills, etc.

Exhibit 4.1: Theoretical generalization about the nature of evaluation and instruction.

4.3 THE TAXONOMY OF EDUCATIONAL OBJECTIVES

A few years later a committee led by Benjamin Bloom including Tyler, met to develop a *Taxonomy of Educational Objectives* based on Tyler's definition of an objective [5]. Present day debates can be traced back to this development.

The "group believed that some common framework used by all college and university examiners could do much to promote the exchange of text materials and ideas for testing. They also believed that such a framework could be useful in stimulating research on examinations and education. After considerable discussion, there was agreement that the framework might be best be obtained through a system of classifying the goals of the educational process using educational objectives" [6]. It certainly stimulated the research and development of the Advanced level examination in Engineering Science in England [7].

The committee distinguished between three domains—the cognitive, affective and psychomotor. The first volume on the cognitive domain was published in 1956. The goals of education were stated in six classes or categories of educational behaviors arranged in order of complexity. In this way they formed a taxonomy. The categories are shown in Exhibit 4.2 (column A). Each of the categories was then sectionalised into sub-categories (e.g., Exhibit 4.2 column B). Each sub-category was further subdivided (e.g., Exhibit 4.2 column C), and finally an expression of what the student would be able to do was reached (e.g., Exhibit 4.2 column D). The committee wrote, "this taxonomy is intended to be a classification of the student behaviors which represent the intended outcomes of the educational process."

In the late 1980's educators began to favor the term outcome instead of objective [8], and there were some specious attempts to differentiate between objectives and outcomes which the group certainly did not intend. The group recognized that the actual behavior of students may differ from that of the intended behavior specified by the objectives.

While Bloom's group believed that the categories were generalizable across the curriculum, they accepted that some subjects might have difficulty in reconciling their requirements with the categories. The categories of geometrical and engineering drawing shown in Exhibit 4.3 illustrate this point. The curriculum authorities felt it necessary to introduce categories for "technique" and "visualization." Those responsible for an engineering science examination offered by the same authority introduced a category for "communication" and, added "design" within the "synthesis" category [9]. Ken Ball of the Engineering Design Research Unit at the University of Liverpool had advised the examining authorities, and they circulated a paper that he had written in a now extinct bur reputable publication (Discovery) on the differences between analysis and synthesis (See Exhibit 4.4). He caused the concept of "originality" to be included and defined in the *Notes for the Guidance of Schools*. At one stage in their *Notes for Guidance* they included a category of creativity [10]. The engineering science examiners did not think the categories were hierarchically ordered.

As stated, the limited requirements for "communication" would not meet the requirements demanded by Trevelyan following his study of expertise among engineers at work [11]. Clearly, all engineering educators have a responsibility in this area.

The Taxonomy became the most cited educational work of the 20th century. It had worldwide impact, and its use is often reported at the major engineering education conferences.

A. Principle Cognitive Domains of the Taxonomy	B. Sub categories of the Domain of Comprehension	C. Domain of Comprehension Sub-category of Translation	D. Examples of Abilities and Skills Related to the Category of Translation
Knowledge Comprehension Application Analysis Synthesis Evaluation	Translation Interpretation Extrapolation	"Comprehension as evidenced by the care and accuracy with which the Communication is paraphrased Or rendered from one language to the Form of communication to another. Translation is judged on the basis of faithfulness and accuracy, that is on the extent to which the material in the original communication is preserved, although the form of the communication has been altered."	The ability to understand non-literal statements (metaphor, symbolism, irony, exaggeration). Skill in translating mathematical verbal material into symbolic statements and vice-versa.

Exhibit 4.2: The Domains of the Taxonomy of Educational Objectives (A) showing the domain of comprehension with its sub-categories, the introduction to the sub-category of translation and two elements of that sub-category. **Currently sub-abilities of the type shown in D are often called learning outcomes.** They are preceded by the phrase "The ability to" and followed by a verb requiring action e.g., to identify; to communicate: to analyse: to make." (Bloom, B. et al. (1964). *The Taxonomy of Educational Objectives*. I. *The Cognitive Domain*. London, Longmans Green).

In 1994 the National Society for the Study of Education published a 40 year retrospective [12]. Many of those who contributed to this volume went on to develop and publish a revised Taxonomy in 2001 [13]. Its categories are listed in Exhibit 4.5. It will be seen that the knowledge category has been substantially changed. Distinctions are made between factual, conceptual, procedural and metacognitive knowledge. Various sub-categories are open to

Knowledge and abilities to be tested.
Knowledge
Technique
Visualization and interpretation
Application
Analysis
Synthesis

Exhibit 4.3: Ability categories to be tested in geometrical and engineering drawing Advanced level (Joint Matriculation Board, Manchester).

criticism, as for example, the illustration given of self-knowledge in the metacognitive category which reads, "knowledge that critiquing essays is a personal strength, whereas writing essays is a personal weakness; awareness of one's own knowledge level."

The cognitive process dimension recognizes "create" as its 6th criteria. The category of "comprehension" is lost, and new categories of "remember" and "understand" are introduced. The inclusion of "understand" is surprising, since one of the reasons for not using it in the earlier *Taxonomy* was that you could not assess understanding. "Understanding" is a word that is commonly used by educators, and it is this fact that is now recognized (p. 269–270). That is the reason for its inclusion in the revised Taxonomy [14]. The separate category of "synthesis" was also omitted.

The skills of analysis, synthesis and evaluation in the original *Taxonomy* are sometimes called Higher Order Thinking Skills (HOTS). They are associated with "critical thinking" and "problem solving" which the authors say are covered by the revised Taxonomy, in so far as they cut across rows, columns, and cells (p. 270). Since the engineer's identity is particularly associated with "problem solving," skill in problem solving is likely to remain a major aim of engineering educators irrespective of the revised Taxonomy. We will consider this issue again in Journeys 5 and 6.

4.4   EISNER'S OBJECTIONS TO THE OBJECTIVES APPROACH

When we speak of learning outcomes in the context of the objectives movement we mean "intended learning outcomes," but as Eisner points out they are often accompanied by "unintended" outcomes that may or not be beneficial to the learner.

Although Eisner accepted the case for pre-formulated goals, he argued that there are many activities for which we do not pre-formulate goals. We undertake the latter in anticipation

Before a designer can apply scientific principles to solve a particular problem he must first understand the scientific principles themselves. At this point the first educational problem arises. The student understands scientific principles by treating them in analytical way: this treatment unavoidably suppresses his ability to handle problems in engineering design, in which the approach is dominated by synthesis.

A typical undergraduate problem illustrates the difference between the analytic and synthetic approaches. Diagram 1 shows an idealized theoretical model of a simple bridge, represented by a thin weightless rod resting on two supports which is subjected to a vertical load. From this information the student can calculate the bending moment distributions across the span. By adding an additional piece of information derived from the thickness and width of the beam forming the span, he can determine the working stresses and compare them to the failure stresses, to predict whether such a bridge could carry the applied load satisfactorily, this approach to structures is analytical in nature since it presupposes that the span, position of loading, type of support, geometric properties of the beam, and the material from which the beam is made are all known factors.

Diagram 1

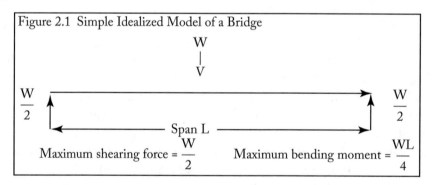

Diagram 2

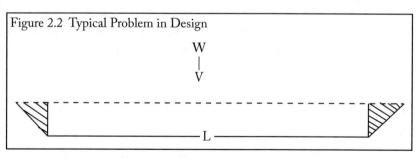

Exhibit 4.4: Extract from K. Ball's article in *Discovery* (April, 1966), Design philosophy in engineering. (*Continues.*)

In contrast, the design problem which had to be solved before the analysis is possible is solved in a very different manner (diagram 2). In the case of the bridge the only known factors are the load which the bridge must carry and the distance, or span, over which the load is to travel. Having introduced a solution before the problem was stated the reader will have already been conditioned to the obvious solution of diagram 1 whilst looking at diagram 2. However, the essence of the real problem is synthesis rather than analysis.

Synthesis, or engineering design in this case, is concerned with the creation of a system which will meet a specified need under conditions where the end product cannot necessarily be foreseen….. the essential difference between an analytical approach to the problem and a synthesised solution. In practice, the engineering of a product must proceed by means of synthesis or creative thinking. This is achieved by a mixture of scientifically based assumptions and estimates concerning a particular part, or sub-system of the overall system, coupled with an analytic investigation of the behavior of the sub-system as the factors controlling it change: the analytical method can then be applied to the sub-system more accurately than before. The procedure is repeated for each sub-system until eventually the understanding of the various sub-systems leads to an understanding of the overall system.

Exhibit 4.4: (*Continued.*) Extract from K. Ball's article in *Discovery* (April, 1966), Design philosophy in engineering.

that something will happen even though we cannot specify what. For example, we do not think much beyond the data, even though we could predict from the ample criteria at our disposal. We evaluate retrospectively what happened against these criteria. Eisner then deduces that teachers should be able to plan activities that do not have any specific objectives. He calls these "expressive activities." They precede rather than follow expressive outcomes. To obtain expressive outcomes a teacher creates activities that are seminal. Eisner writes, "What one is seeking is to have students engage in activities that are sufficiently rich to allow for a wide, productive range of valuable outcomes."

Eisner would probably have said to those who would provide well-defined outcomes for research based final year projects, leave them alone for detailed assessment taxonomies might limit expression. There is a need to examine the effect of long lists of outcomes on general learning and creativity, in particular the strategies that students adapt to meet the requirements of assessment, and to examine the impact of accidental competencies on learning [15].

Related to Eisner's objections is the effect *The Taxonomy* had on the language of teaching. It removed terms such as understanding, critical thinking, and motivation that were part of the teacher's emotional vocabulary. To some extent the revisers recognized that this was a problem and introduced the category of "understand." But they did not categorize "problem solving" or "critical thinking" which they felt were covered by other sub- categories.

A. Knowledge	B. Examples of Sub-classification	C. Categories and Cognitive Processes	D. Examples of Sub-categories
Factual knowledge		Remember	
Conceptual knowledge		Understand	Interpreting (Clarifying, paraphrasing, representing, translating).
Procedural knowledge			
	Knowledge of subject-specific skills and algorithms		Exemplifying (Illustrating, instantiating).
	Knowledge of subject-specific techniques and methods		Classifying (categorizing, subsuming).
	Knowledge of criteria for determining when to use appropriate procedures		Summarizing (abstracting, generalizing).
			Inferring (concluding, extrapolating, interpolating, predicting).
Metacognitive knowledge			Comparing (contrasting, mapping, matching).
		Apply	Explaining (constructing models).
		Analyze	
		Evaluate	
		Create	
			Generating (hypothesizing)
			Planning (designing)
			Producing (constructing)

Exhibit 4.5: From the summary of the revised Taxonomy. Anderson, L. W. and D. R. Krathwohl (2001). *A Taxonomy for Learning, Teaching, and Assessing. A revision of Bloom's Taxonomy of Educational Objectives*. New York. Addison Wesley Longman (Abridged edition). The summary of the categories and cognitive process dimension contains a separate list of alternate words for the sub-category title. Examples are given in brackets in column D.

Given the criticisms of *The Taxonomy*, several of which have not been considered here [16], it is not surprising that other Taxonomies have been developed that have interested engineering educators [17].

In engineering the terms "critical thinking" and "problem solving" contribute to the way of thinking in engineering. Attempts have been made to describe the abilities that contribute to such skills. Of somewhat more importance than the issue of language is the overwhelming influence of utilitarianism on educational policy, to the extent that reasoned argument about the aims of education from other perspectives has virtually ceased.

4.5   INSTRUCTIONAL PLANNING

For some years after the publication of *The Taxonomy* the term "behavioral objective" was used. Some authorities continue to use it. For example, in 1977, in the United Kingdom Cohen and Mannion in their best-selling book on teaching practice, suggested that students should, at the beginning of planning a lesson(s), state its (their) aim(s), non-behavioral objective(s) and behavioral objective(s). "Aim" in this sense is something much more limited than those that are discussed by philosophers [18].

I confess that both my students and I often found it difficult to distinguish between the aim and the non-behavioral objective. In 2013, two engineering educators lost the distinction between aim and non-behavioral objective with the substitution of "goal." for both terms. They also substituted "learning outcomes" for "behavioral objectives." Exhibit 4.6 shows how I used Cohen and Mannion's advice to design two lessons on the construction of objective items and procedures for objective test analysis. The behavioral objectives specify what should be tested.

The first direct attempt to describe the value of *The Taxonomy* to engineering educators seems to have been due to Jim Stice in what must be regarded as a seminal paper in *Engineering Education* [19]. He arrives at *The Taxonomy* via a discussion of "instructional objectives" which had become important for the developers of programmed instruction. He followed the advice of Mager [20] who had written a much read book on the topic. Mager had defined an objective as an "intent communicated by a statement describing a proposed change in the learner." Nowadays, the literature drops the intended and simply uses *learning outcome*.

Mager evidently thinks that the student demonstrates a competence because his instruction reads "describe the important conditions under which the learner will demonstrate competence." How students learn a competence is a major issue for beginning engineering educators, and relates to their understanding of learning more generally, matters which will be taken up in later chapters. Holdhusen, James-Byrnes and Rodriguez have described a lesson study for a distance education statics course. The idea is that faculty should develop, teach, observe, analyse, and revise a single lesson for a single class period. "During the planning, the instructors anticipate student, reactions, interpretations, and difficulties with the lesson and alter the instructional experiences accordingly […] One instructor delivers the lesson while the other instructors observe student learning. The group then analyses these observations and the lesson is revised" [21].

Aim
To introduce the principles of objective testing (2 lectures)

Non-Behavioral Objective or Goal
At the end of three classes the student will be introduced to the principle types of objective item, how they are written, and how objective test results are analyzed

Behavioral Objectives or Learning Outcomes or Competencies.
At the end of the exercise the student will be able to:
  (a)   Construct objective items into a test in the subject taught by the student.
  (b)   Conduct a short classroom test.
  (c)   Recognize the limitations of such tests.
  (d)   Conduct an item analysis of the test.
  (e)   Evaluate the analysis and suggest changes in the items where this is thought necessary.

Exhibit 4.6: Cohen and Mannion's advice was used to design two lessons on the construction of objective items and procedures for objective test analysis.

Jim Stice reported that when he adopted an objectives approach to the design of his courses, it made him ask questions about what was essential and what was not essential. He found that some of what he had been teaching was inessential. This suggests that a really thorough going analysis of a curriculum by objectives might lead to a reduction in content and, therefore, reduce the load on students. This means that the syllabus should not be fitted to the objectives, but derived from an analysis of the time required to learn key objectives.

4.6   QUESTIONING, QUESTIONS, AND CLASSROOM MANAGEMENT

Classroom observations suggest that up to 80% of classroom time may be used for questioning, and that most of the questions belong to the knowledge and comprehension categories of the original taxonomy. Thus, memory skills are encouraged at the expense of higher order thinking skills [22]. Of the several reasons for this state of affairs three may be singled out.

First, it may be due to lack of training in the design of questions for oral and written use. Even experienced teachers of engineering have a fear of asking what have come to be called "wicked questions." Some teachers have a fear of scoring responses to questions where there is no right answer, where student judgments have to be assessed. Exhibit 4.7 [23] shows how command words and question type can help engineering educators design questions. To get beyond level 3 of *The Taxonomy* demands that teachers stand outside the box, as well as engage in dialogue with their colleagues, in order to learn how to develop such questions [24], and there are reports that discuss the design of complex questions [25], and the ordering of questions to achieve greater in-depth understanding [26].

Taxonomy Category	Command Words and Question Type
Knowledge	Arrange, define, describe, match, order, memorize, Name, note, order, repeat Who? What? When?, Where? Questions.
Comprehension	Alter, change, clarify, define in your own words, Discuss, explain, extend. Give examples, translate.
Application	Apply, calculate, compute, construct, operate Practice, How many? Which? Write an example question.
Analysis	Analyze, appraise, categorize, compare, conclude Contrast, criticize, diagnose, differentiate, etc. Why? Questions.
Synthesis	Assemble, compile, compose, create, improve, Synthesize, what if\? How can we improve? What would happen if? How can we solve? Questions.
Evaluate	Appraise, argue, choose, certify, criticize, decide, Deduce, defend, discriminate, estimate, evaluate, Recommend, etc.

Exhibit 4.7: The categories of the *Taxonomy of Educational Objectives* (original version) showing command words that begin statements and relevant question type (After Batanov, D. M., Dimmit, N. J. and W. Chookittikul (2000). Q and A teaching/learning model as a new basis for developing educational software. *ASEE/IEEE Proceedings Frontiers in Education Conference*, 1, F2b–12 to 17.

It is equally important to teach students to ask questions, that is the right questions for the ability to ask questions is at the heart of the competency of "diagnosis." In fact, questioning is at the heart of the engineering process as will be seen from Trevelyan's analysis of the expert engineer [27].

Finally, even the most experienced teacher experiences lessons that go badly wrong. A catalogue of such experiences with ideas of how to undertake damage control when they do has been made by Gehringer [28].

4.7   RECONCILIATION: A CONCLUSION

Discussion of *The Taxonomy* for the cognitive domain has completely overshadowed the publication of the volume on the affective domain. Engineering educators are beginning to recognize that this domain, however, defined is of considerable importance in engineering practice [29, see Chapter XIV].

The objectives movement has made us aware that we should be clear about what it is we are trying to do. The problem is that in the search for validity and reliability it has become so reductionist that there is a danger that long lists of outcomes, many of them trivial, cause a search for "ticks" rather than learning. But, *The Taxonomy* has shown us how to build up our own categories and validate them. Among those of importance to engineers are problem solving, problem finding, decision making, design, and diagnosis which will be considered in the next chapter.

In the *Assessment of College Performance* (1979) [30], R. I. Miller tried to reconcile the behavioral objectives approach with that of its opponents. He wrote, "(1) Objectives expressed in measurable, behavioral terms are appropriate for basic skills and for other areas where there is agreement about the components of an instructional program. (2) For most purposes, behavioral objectives need not be reduced to trivial detail. The degree of specificity may vary and should relate to the purpose of instruction and the understanding of students and instructors. (3) The use of behaviorally stated objectives should be contained in an instructional model which recognizes and provides for individual differences. (4) Complex and long-range objectives should be included in a set of objectives even though they cannot be described in precise terms or measured with a high degree of accuracy. (5) Educational objectives must be appropriate to the social milieu at a given time, and students should participate with their instructors in finding objectives that make sense to them. (6) In times like the present when technological and social changes are rapid and the future uncertain, the desired behaviors should be adaptable to situations other than the existing one. The ultimate usefulness of behavioral objectives will depend on how effectively they may be adapted to quite different learning needs and situations." All of which seems to allow for expressive and focussing objectives [31], as well as competencies that are accidental.

NOTES AND REFERENCES

[1] Tyler, R. (1949). *Basic Principles of Curriculum and Instruction*. Chicago. Chicago University Press. 39

[2] Culver, R. S. and J. T. Hackos (1982). Perry's model of intellectual development. *Engineering Education*, 73, (2), pp. 221–226. 39

[3] See Chapter 7 Heywood, J. (2005). *Engineering Education. Research and Development in Curriculum and Instruction*. Hoboken, NJ, IEEE/Wiley. 40

[4] Rodriguez-Marek, E. (2017). Connecting the data in assessment from course student learning objectives to educational program outcomes to ABET assessment. *Proceedings Annual Conference of the American Society for Engineering Education*. Paper 316. 40

[5] *loc. cit.* 40

[6] Bloom, B. S. (1994). Reflection on the development and use of the taxonomy in L. W. Anderson and L. A. Sosniak (Eds.), *Bloom's Taxonomy. A Forty Year Retrospective*. National Society for the Study of Education. Chicago. Chicago University Press. 41

[7] See appendix B of Heywood, J. (2016). *The Assessment of Learning in Engineering Education. Practice and Policy*. Hoboken, NJ, IEEE/Wiley. 41

[8] Otter, S. (1992). *Learning Outcomes in Higher Education*. London. HMSO for Employment Department. 41

[9] Carter, G., Heywood, J., and D. T. Kelly (1986). *Case Study of Curriculum Assessment. GCE Engineering Science (Advanced)*. Manchester. Roundthorn Press. 41

[10] The kind of thinking that went on in the committee is illustrated by the following statement that I wrote in 1972 (i). "An approach to a definition of creativity arises from its common usage in respect of individuals. 'Creative is the label we apply to the products of another person's originality' (ii). This is with improvements in originality defined as a behaviour 'which occurs relatively infrequently, is uncommon under given conditions, and is relevant to those conditions' (iii). The terms in this sense were used by Pelz and Andrews (iv) in their study of scientists in organizations. As Macdonald says, 'it is relatively easy to translate this definition into behavioural specifications, and it places originality in the model of problem solving, obviating some of the difficulties but, one should go further than Macdonald and point out that the need for external judgement implies a relationship between the person and the judge, which leaves the person as agent in the process. This is to follow Macmurray's philosophy' (v). The impact of an organization, society, call them what one will, play a major role in, first, determining what is creative, and second, in allowing its fulfilment. As Vernon (vi) says, the influence of home, leisure

and education on individuals may be profound. Before looking at 'abilities' and 'problem solving' one must look first at the role of the organization in creativity." 41

(i) Heywood, J. (1972). Short courses in the development of creativity in S. A. Gregory (Ed.), *Creativity and Innovation in Engineering*. London, Butterworths.

(ii) Macdonald, F. J. (1968). *Educational Psychology*. Belmont, CA. Wadsworth.

(iii) Maltzmann, J. et al. (1960). Experimental studies in the training of originality. *Psychological Monographs*, 493.

(iv) Pelz, D. C. and F. M. Andrews (1966). *Scientists in Organizations. Productive Climates for R & D*. New York. Wiley.

(v) Macmurray, J. (1957). *The Self as Agent*. London. Faber and Faber.

(vi) Vernon, P. E. (1964). Creativity and intelligence. *Educational Research*, 6, 163.

[11] Trevelyan, J. (2014). *The Making of an Expert Engineer*. London. Taylor and Francis, (CRC Press). 41, 55

[12] L. W. Anderson and L. A. Sosniak (Eds.), *Bloom's Taxonomy. A Forty Year Retrospective*. National Society for the Study of Education. Chicago. Chicago University Press. 42

[13] Anderson, L. W., and D. R. Krathwohl, et al. (2001). *A Taxonomy for Learning, Teaching, and Assessing. A Revision of Bloom's Taxonomy of Educational Objectives*. New York, Addison Wesley Longman. 42

[14] *ibid.* "Discussion of the Handbook in the years since its development has made it clear that teachers miss having a place where the term 'understand' can 'fit.' The result is that, in determining how best to construct our framework, we considered a different criterion-namely, that the framework should embrace the terms that teachers frequently use in talking about education. We replaced 'comprehension' with 'understand' simply because the group working on this volume gave more weight to the universal usage of the term in selecting names for the categories" (p. 269). 43

[15] The idea of accidental competence is due to Walther and Radcliffe. It has affinities with Eisner's expressive outcomes and the hidden curriculum. They are attributes that are not achieved through targeted instruction. They are acquired through unintentional coactions of curricular elements or aspects surrounding the educational process. (Walther, J. and D. Radcliffe (2006). Engineering education: targeted learning outcomes or accidental competencies. *Proceedings Annual Conference of the American Society for Engineering Education*, paper 1889). 45

[16] See Chapter 2 of Heywood, J. (2005). *Engineering Education. Research and Development in Curriculum and Instruction*. Hoboken NJ, IEEE/Wiley. 47

[17] The SOLO Taxonomy is of interest (i). It attempts to link the forms of knowledge with development. There are five modes of learning which have many similarities with Pi-agetian/Brunerian development stages. These are sensori-motor, ikonic, concrete symbolic, formal and post formal. The forms of knowledge related to these are tacit, intuitive, declarative, theoretical and meta theoretical. There are five structural levels (hierarchically ordered) in a learning cycle that is repeated in each of the forms. 47, 56

Both Gibbs (ii) and Ramsden (iii) have described these levels in terms of the type of answers a person might give to a question. In this way it might be related to degree levels, but this has been questioned by Imrie (iv). Following Gibbs the levels are:

1. Pre-structural. A stage of ignorance where the learner has no knowledge of the question.

2. Unstructural. Where the learner is able to give answer that contain one correct feature.

3. Multistructural. Where the answer contains a check list of answers.

4. Relational. The answer integrates the items into an integrated whole.

5. Extended abstract. The answer is related to the more general body of knowledge.

Its levels have also been related to deep and surface learning as well as the design of examination questions (v).

It is of interest to compare these levels with the way the rubrics for coursework assessment are structured in Engineering Science at A level (vi).

(i) Biggs, J. B. and K. F. Collis (1982). *Evaluating the Quality of Learning. The Solo Taxonomy.* New York. Academic Press.

(ii) Gibbs, G. (1982). Improving the quality of student learning through course design in R. Barnett (Ed.), *Learning to Effect.* Buckingham. SRHE/Open University Press.

(iii) Ramsden, P. (1992). *Learning and Teaching in Higher Education.* London. Routledge.

(iv) Imrie, B. W. (1995). Assessment for learning. Qualities and taxonomies. *Assessment and Evaluation in Higher Education*, 20(2), pp. 175–189.

(v) Baillie, C. and P. Walker (1998). Fostering creative thinking in student engineers. *European Journal of Engineering Education*, 23(1), pp. 35–44.

(vi) See chapter 3 of Heywood, J. (2016). *The Assessment of Learning in Engineering Education. Practice and Policy.* Hoboken, NJ, IEEE/Wiley.

Fink's taxonomy cuts across the cognitive and affective domains. Its principle categories are Foundational knowledge: application: integration: human dimensions: caring: learning how to learn. For an example of its application in engineering see Ferro, P. (2011). Use of Fink's taxonomy in establishing course objectives for a redesigned materials engineering course. *Proceedings Annual Conference of the American Society for Engineering Education.* Paper 309.

[18] Cohen, L. and L. Mannion (1977). *A Guide to Teaching Practice.* (1st ed.), London, Methuen. 47

[19] Stice. J. E. (1976). A first step toward teaching. *Engineering Education*, 67, pp. 394–398. 47

[20] Mager, R. F. (1962). *Preparing Instructional Objectives.* New York. Fearon Publishers. 47, 75

[21] Holdhusen, M., James-Byrnes, C., and L. Rodriguez (2017). Lesson study for a distance education statics course. *Proceedings Annual Conference of the American Society for Engineering Education.* Paper 741. 47

[22] Bellon, J. J., Bellon, E. C., and M. A. Blank (1992). *Teaching from a Research Knowledge Base—A Development and Renewal Process.* New York. Merrill/Macmillan. 48

[23] Batanov, D. M., Dimmit, N. J., and W. Chookittikul (2000). Q and A teaching/learning model as a new basis for developing educational software. *ASEE/IEEE Proceedings Frontiers in Education Conference*, 1, F2B-12 to 17. 48

[24] For example John Cowan argued that conventional engineering examinations as set in the UK (e.g., calculate the position of the centroid of this section, and of its second moment of area about a horizontal axis through the centroid—accompanied by a diagram) "call for the student to know a procedure, or an algorithm, and to understand it sufficiently well to be able to apply it in situations which are often familiar ones. Thus he is required to operate somewhere between the second and third levels on the much-maligned Bloom scale of cognitive ability." 49

Cowan suggested that if the level was to be increased it would be necessary to think outside of the box of question design. He had in other publications drawn attention to the value of quantitative thinking in engineering. So he re-designed the question to read "Without calculating or using total values of second moment of area, arrange the following sections in ascending order of magnitude of second moment of area and give sufficient reason to justify your decisions—(diagram 4.8)." He called these ascending order questions and reported of these first year examples. He argued that for a student to answer a question in the second form without first comparing and thereby analysing the various diagrams, thus applying his knowledge and understanding of the principles to be tested by the question. He therefore, operates, in the critical opening stages of this problem solving, at the fourth level of the Bloom scale.

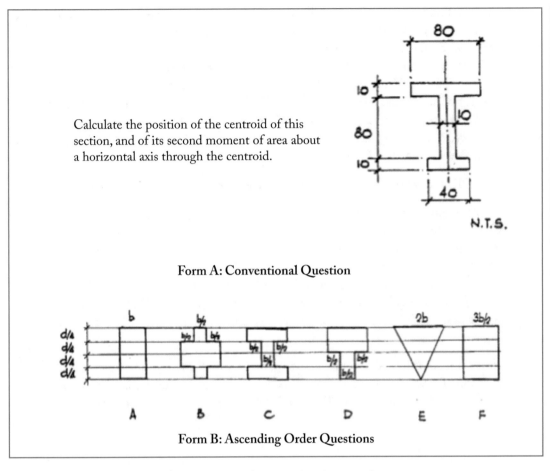

Calculate the position of the centroid of this section, and of its second moment of area about a horizontal axis through the centroid.

Form A: Conventional Question

Form B: Ascending Order Questions

Exhibit 4.8: Cowan's non-traditional approach to the design of examination questions to test qualitative understanding—Forms A and B.

[25] Brophy, S. (2017). First year engineering students' initial ideas for solving complex problems. *Proceedings Annual Conference of the American Society for Engineering Education*. Paper 2498. 49

[26] Das, N. (2017). Interactive tutorial modules for basic mechanics topics. *Proceedings Annual Conference of the American Society for Engineering Education*. Paper 928. 49

[27] *loc. cit.* note [11]. 50

[28] Gehringer, E. (2017). Damage control: What to do when things don't work. *Proceedings Annual Conference of the American Society for Engineering Education*. Paper 1483. 50

[29] The idea that professional/soft skills cannot be developed without attention to the affective is beginning to be understood and several studies in engineering education support this view. Some of these are recorded in Heywood, J. (2016). *The Assessment of Learning in Engineering Education.* Hoboken. NJ, IEEE/Wiley. See the taxonomies developed in studies of steel making firms by W. Humble, p. 99. See Ferro note [17]. In relation to the learning of mathematics see Goold, E. and F. Devitt (2014). Mathematics in engineering practice; tacit trumps tangible in B. Williams, J. Figueiredo, and J. Trevelyan (Eds.), *Engineering Practice in a Global Context. Understanding the Technical and the Social.* London CRC Press/Taylor and Francis. 50

[30] Miller, R. I. (1979). *The Assessment of College Performance.* San Fransisco, Jossey Bass. 50

[31] Focussing objectives is a term used by Heywood (1989) who argued that it was only possible for faculty to consider (assess) a limited number of domains for which reason he sometimes called them focussing objectives. The domains selected have to have a high level of significance for the ongoing purposes of a student's education. Focussing objectives are derived from key learning skills that a person will need in the job for which he is trained. Necessarily they incorporate values. They are process skills that lead to the product. In later work he refers to them as ability domains following the practice of Alverno College. 50

Heywood, J. (1989). Problems in the evaluation of focussing objectives and their implications for the design of systems models of the curriculum with special reference to comprehensive examinations. *ASEE/IEEE Proceedings Frontiers in Education.*

JOURNEY 5

Problem Solving, Its Teaching, and the Curriculum Process

5.1 INTRODUCTION

There are significant differences between those who believe that problem solving should be taught, and those who do not believe it to be a problem because problem solving skills develop as the learner is totally immersed in the subject [1]. The former would seem to be representative of the social efficiency ideology, and, it would seem, the latter belong to the scholar academic ideology. In the view of the former knowledge is a skill which can be learned. It is to quote a distinguished psychologist of the last century, Robert Gagné, identified by the "successful performance of tasks" [2]. His model will be considered in Journey 10.

There is a sub-plot between those who believe that problem solving can be taught within normal course structures and those who believe it should be taught in separate courses. Two courses that aim to teach thinking skills, and have had some success are the instrumental enrichment [3], and philosophy for young children programmes [4]. Both Karl Smith and Russell Korte [5] and Bill Grimson [6] have demonstrated the value of philosophy and its methods of reasoning in engineering.

In engineering, the former position may be represented by W. E. Red [7] whose approach will be discussed in Journey 6. The best known example of the latter is the McMaster Problem Solving Course developed by Woods and his colleagues [8]. It needs to be stressed that problem solving courses in engineering are designed to achieve knowledge and skill goals. Courses that are designed around problems are often called Problem Based Learning (PBL). They can also be designed around projects [9].

There are two extensive reviews of problem solving in engineering education. They approach the topic in quite different ways. A 2005 review considered how engineering educators had responded to the need to teach problem solving skills [10]. In a review in 2014, Jonassen reviewed research from the perspective of problem solving in the work place [11](a), and also the specific problem of "transfer," so necessary in solving industrial problems, in physics [11](b).

In this journey and Journey 6, the focus is on what educators have revealed about their approaches to teaching problem solving, decision making and critical thinking that may be helpful. They will also be examined for the contribution they may make to the construction of a category of "problem solving." It should be noted that lists of sub-abilities that contribute to problem solving and critical thinking had been developed in the 1960's and 1970's in both the school and higher education systems (Exhibit 5.1) [12, 13]. In Journey 7 these issues will be considered from the perspective of intelligence and the question "Can we teach intelligence?"

The Critical Thinker

1. Asks significant and pertinent questions and states problems with specificity. Arrives at solutions through hypothesis inquiry, analysis, and interpretations.

2. Assesses statements, insights and arguments according to the knowledge and skills provided by formal and informal logic and by the principles of aesthetic judgement.

3. Derives meaning through an educated perception, whether propositional, systematic or intuitive.

4. Formulates propositions or judgements in terms of clearly defined sets of criteria.

5. Strives to acquire knowledge of the various disciplines, knowing that such knowledge is a necessary, though not sufficient condition for critical thinking.

6. Understands the different modes of thought appropriate to the various disciplines. Can apply these modes of thought to other disciplines and to life.

7. Is aware of the context setting in which judgements are made, and of the practical consequences and values involved.

8. Thinks about the world through theories, assessing these theories and their contexts to determine the validity of their claims to knowledge of reality.

9. Seeks and expects to find different meanings simultaneously present in a work or event. Is intrigued and curious about phenomena others might avoid, disavow, or disagree.

10. Recognizes and accepts contradiction and ambiguity, understanding that they are an integral part of thought and creativity.

11. Constructs and interprets reality with a holistic and dialectical perspective. Sees the interconnectedness within a system and between systems.

12. Is aware of the problematical and ambiguous character of reality. Understands that language and knowledge are already interpretations of reality.

13. Tolerates ambiguity, yet can assume a committed position.

14. Is aware of the limitations of knowledge and figures epistemological humility.

Exhibit 5.1: Cromwell's profile of the critical thinker in the arts and humanities [8].

This chapter is primarily concerned with problem solving, the curriculum process and assessment.

5.2 DEFINITIONS AND APPROACHES TO TEACHING PROBLEM SOLVING

One of the problems in this area of study is that linear models of problem solving, critical thinking and decision making are almost always defined by categories of problem finding (formulation), problem analysis (alternative solutions, assumptions made), plan of action, actions, outcome, and evaluation (feedback), yet the literature produces quite different insights. Here the concern is with problem solving, but with two caveats.

First, F. J. McDonald asserted that problem finding/problem formulation is a different competency to the competency of problem solving [14]. (I use competency to describe a grouping of sub-abilities that make up a particular skill.) Second, two Scandinavian's Sutinen and Tarhio distinguished between problem solving and problem management.....“the term ‘management’ instead of ‘solving’ (problem management) stresses that a problem always undergoes a process. A solution of a problem is nothing more than one of the stages of this process: a potential end-product to be evaluated before finishing, the more extensive process” [15]. This view is in keeping with the axiom that a learner will be better able to learn to solve problems if the learner has a model of the problem solving process in mind such as that suggested by J. L. Saupé [16] (see Exhibit 5.2). The models that some engineering educators have used to teach problem solving skills are clearly heuristics of the process.

1. Ability to recognize the existence of a problem.	(Problem finding)
2. Ability to define the problem.	(Problem solving)
3. Ability to select information pertinent to the problem.	(Problem solving)
4. Ability to recognize assumptions bearing on the problem.	(Problem solving)
5. Ability to make relevant hypotheses.	(Problem solving)
6. Ability to draw conclusions validly from assumptions, hypotheses and pertinent information.	(Problem solving)
7. Ability to judge the validity of the processes leading to the conclusion	(Problem solving)
8. Ability to evaluate a conclusion in terms of its assessment.	(Problem solving)

Exhibit 5.2: J. Saupé's steps in the process of problem solving [11].

Crudely speaking there have been three approaches to the development of problem solving skills as alternatives to total immersion. The first is to design assessments that test problem finding and problem solving skills [17, 18]. The second is based on the use of heuristics as advocated by Billy Koen [19]. He claims that this is a universal method that extends beyond engineering [20]. The third stems from trying to find out how engineers learn. It is in the tradition of that research which tries to establish the differences between experts and novices pioneered by such persons as the Nobel Laureate Herbert Simon [21].

5.3   TYPES OF PROBLEM, DIFFICULTY, AND COMPLEXITY

Kahney distinguished between well-defined problems and ill-structured problems [22]. The latter seem now to be called "wicked" problems. Bolton and Ross [23] who cited Kahney also cited Thompson [24]. He distinguished between open and closed problems: they seem to be at the extreme ends of a spectrum. Closed problems that are amenable to a single correct solution are at one end, and open problems that are not amenable to such a solution are at the other end. Any problem on the spectrum should be able to be taught in the classroom although the time taken for such teaching is likely to vary not only with the complexity of the problem, but with the ability of the students to understand the concepts involved. The complaint against engineering educators is that they do not pay enough attention to the wicked problems that arise in engineering practice.

In many ways the detail of this spectrum can be seen in a table by Dean and Plants which suggests five stages of problem solving sophistication (Exhibit 5.3) [25]. They show a relationship between routines and the higher-level skills of problem solving, as well as the importance of routines in the development of open-ended problem solving skills.

Stages of Problem Solving Sophistication

Routines

Operations which, once begun, afford no opportunity for decision, but proceed by simple or complex mathematical steps to a unique solution.

Diagnosis

Sorting out correct routines from incorrect routines for the solution of a particular problem.

Strategy

The choice of a particular routine for the solution of a problem which may be solved by several routines or variations of routines, all of which are known to the student.

Interpretation

The reduction of a real-world situation to data which can be used in a routine, and the expansion of a problem solution to determine its implications in the real world.

Generation

The development of routines which are new to the problem solver.

Exhibit 5.3: Stages of problem solving sophistication due to R. H. Dean and H. L. Plants [18] and summarized in this form by W. E. Red [3].

Related to this table are five levels of difficulty proposed by D. K. Apple and his associates (Exhibit 5.4) [26]. Inspection of these levels suggests a crude relationship with the levels in the original version of *The Taxonomy of Educational Objectives*.

Level	Description
1. Automatic	Performance of task without thinking.
2. Skill exercise	Consciously involved but minimal challenge using specific knowledge.
3. Problem solving	Challenging, but possible with current knowledge and skills through a strong problem solving approach.
4. Research	Requires additional knowledge that currently does not exist within a learning effort to effectively accomplish the task.
5. Overwhelming	Cannot be accomplished without a significant increase in capacity, most likely to bringing in additional expertise.

Exhibit 5.4: Levels of difficulty in problem solving situations due to D. K. Apple and colleagues [20].

A more specific relationship with the *Taxonomy of Educational Objectives* has been made by Prince and Hoyt [27]. They distinguish between introductory problem solving in which knowledge, comprehension and application come into play: intermediate problem solving involving analysis: and, advanced problem solving involving synthesis and evaluation. Prince and Hoyt make the point that traditional engineering courses rely heavily on textbook problems that do not require problem solving skills that are relevant. The textbook exercises test the material in the chapter being read. They wrote, "That is not the point of 'problem solving' in any real sense." Students, Prince and Hoyt said, while being able to solve textbook problems may not be able to apply the concepts to real problems. This is the problem of "transfer." Skill in transfer can only be developed by showing students how to solve actual engineering problems; the need for carefully chosen exemplars cannot be overstressed (see also chapter X). This is particularly the case in problem based (PBL) or project based learning [8].

Much attention is being giving to the design and solution of "real world" problems. A study by Strobel and Cardella showed that the experience of engineers in the real world of problems are that many of them are compound [29]. That is, they contain a variety of different problem types, moreover, transitions from one type of problem type to another within a compound problem are a unique class of problem themselves.

Clearly, assessment, its structure and design impact on the way students learn problem solving competencies and develop the skills of transfer.

5.4   ASSESSMENT, INSTRUCTION, AND OBJECTIVES–THE CURRICULUM PROCESS

A group of mathematics teachers were asked to take part in the redesign of a public examination in mathematics for fifteen year olds [28]. While they were able to develop objectives for the course they were not allowed to alter the syllabus (content). They were required to trial their experimental examination on the students they taught. This had to be done as part of their regular teaching for the public examination.

A widely held view among these teachers was that the official public examination tested knowledge and comprehension, and that recall played a large role in performance. They wanted to test the higher levels of thinking in particular "analysis" which they associated with problem solving. Surprisingly, when this particular sub-test was evaluated, it was found that the students performed very badly.

Although the reasons given for this result are complex, the evaluators found that because the teachers had to follow the official syllabus, they did not give the pupils specific instruction in the problem solving skills they were demanding, since the syllabus did not demand such skills. This gives support to the view that problem solving requires teaching, and that would, in any event, immerse the students in the subject [30].

Similarly, in a public examination in Engineering Science for 17/18 year olds in England the students were required to complete a substantial project. Before they could begin work on the project they had to complete a formal project proposal. The candidate had to propose a title that was clear and unambiguous, prepare an analysis of the problem, consider the practical problems to be solved, offer possible solutions, state the resources that would be required, and present a timetable for completion [31]. The final report would be a substantial expansion of the proposal together with full details of what had been accomplished. In this way, candidates would be able to demonstrate their capability in applying their engineering science to practical problems, and gain an understanding of the design process.

The examiners decided that they would test the skills involved in project planning in a sub-written test (see Exhibit 5.5). They expected there would be a transfer of skill from the experience of completing the project to completing a project proposal in a written examination: therefore, a high correlation between the marks for coursework assessment and the written paper was predicted. Not so. The evaluation was completed on three successive occasions (including the sub-paper shown in Exhibit 5.5). As in the previous example a number of explanations were possible including criticism of the method of analysis. But among the explanations was the finding that no specific instruction was given in design, or on how to approach the written paper. The need for instruction becomes more apparent if the reason as to why this correlation was found which was offered several years later is taken into account. It suggested that because the times required to complete the two exercises were substantially different (50 hours for the project; 1 hour for the exam) that different kinds of executive skills were required [32].

Section C. **Project Design**

Answer all parts of this question in the twelve-page answer book provided.

Time allowed 1 hour.

This question forms a project planning exercise similar to that undertaken as part of the Course Work requirement. Credit will be given for depth of thought, consideration of practical alternatives where relevant, and for clear statements of reasons for decisions and choices.

You are advised to spend at least 40 minutes on part (b).

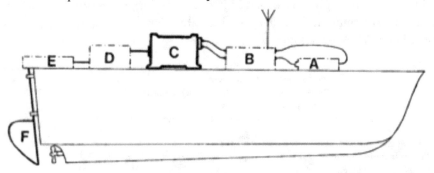

A Battery
B Control unit
C Motor
D Reduction gearing
E Steering mechanism
F Rudder

A schematic diagram showing the place of the motor in the steering-control system.

The rudder mechanism of a remotely controlled model boat, figure, is to be driven by a small electric motor. A range of small motors is available all designed to be operated from a 1.5 volt battery.

(a) List the motor characteristics which need to be known before a motor can be chosen suitable for a given rudder mechanism, reduction gearing and battery system. Indicate the relative importance of these characteristics. Give reasons for your answers.

(b) Design an experimental procedure to determine the most important electromechanical characteristics of the motor. Pay particular attention to the selection and design of the equipment you would use, stating reasons for the decisions made. Outline the procedure you would adopt and show how you would use your observations to enable you to obtain the desired information.

(c) Estimate the total time you would need for the investigation. Draw up a timetable indicating how many hours would be required for the major activities to be undertaken.

Exhibit 5.5: Example of the sub-test in Project Design in the Engineering Science Advanced level Examination set by the Joint Matriculation Board, Manchester, 1973.

These two examples support the view that students need instruction in the knowledge and skills that are to be tested. There is nothing wrong with teaching to the test provided that the test questions are properly designed, and are not tests of recall.

These examples also show the complexity of the curriculum process for which reason I prefer to illustrate it in the non-linear form shown in Exhibit 5.6. You will only achieve outcomes by chance if instruction is not provided on the learning required to obtain them. Recently in the United States engineering educators have begun to call this linkage "alignment." However, "alignment" does not convey the complexity of the problem because if the learning strategy is changed and more time given to achieve effective learning, other outcomes will have to be dropped, and the curriculum will be changed. The curriculum is a dynamic process. The learning of concepts will be discussed in Journey 11.

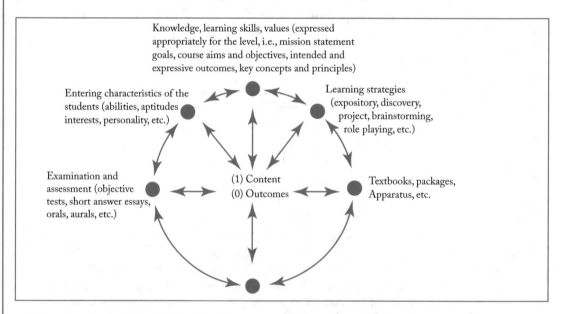

Exhibit 5.6: A model of the curriculum process to indicate (1) the first phase in which the structure of the syllabus content is derived, and (2) how the intended learning outcomes are a function of a complex interaction between all the parameters, and allowing that there will also be unintended outcomes.

As long ago as 1976 a group of freshman students wrote about their experience and gave their view of the factors that contribute to effective problem solving in *Engineering Education* [33]. These were:

1. There must be a problem or an awareness that a problem exists.

2. Six pre-requisite skills and attitudes are essential. These are:

(i) The basic knowledge pertinent to the problem(s).

(ii) The learning skills necessary to obtain the information necessary to solve the problem.

(iii) The motivation to want to solve the problem.

(iv) The memorized experience of factors that provide order of magnitude "feelings" as to what assumptions can be made and how reasonable the answer is.

(v) The ability to communicate the answer; and perhaps,

(vi) Group skills if the problem must be solved by a group of people.

3. An overall organized strategy is required.

4. For specific steps in the strategy, there are well known alternatives.

5. A problem solver uses four abilities time and time again. These are to create, analyse, generalize, and simplify.

6. Sets of "good hints" or "heuristics" have been developed about what to do next.

This is a useful guide for beginning educators provided that it is understood that since 1976 numerous "good hints" and "heuristics" have been proffered in the literature, and advances have been made in our understanding of problem solving, and the needs that students of different levels of learning have. The students make no reference to the time needed for learning, or the learning level at which a problem should be set.

5.5   DIFFICULTY IN, AND TIME FOR LEARNING

The evaluators of the mathematics teacher's designed examination papers referred to previously, found that too often the teachers designed questions that were too difficult for the students. There were a number of reasons as to why this should have been, as for example, the relative aptitudes of the pupils tested. Nevertheless, it seems that mathematics teachers are not the only ones that do, or have done this. I plead guilty! The difficulties that some instructors have with freshman students may be that they over-estimate their specific aptitudes in relation to the subject matter. They may also underestimate the time that students require to comprehend a given concept. These are matters to which we will return in Journey 10. In the meantime it is instructive to consider how experts differ from novices.

One way of distinguishing novices from experts is to get them to draw concept maps of how they perceive a particular problem. In one study in biomedical engineering. Walker and King [34] reported that concept maps of a traditional kind revealed that when experts and novices were asked to illustrate the relationships between the 10–20 most important concepts in biomedical engineering, the expert's maps were much more dense than the maps of the novices. They conducted a second study which obtained concept maps at different times during the course and found that the later maps had more concepts, more precise vocabulary, and greater validity.

In civil engineering Fordyce [35] used an unconventional approach to mapping the concept of stress to try and understand the cognitive structure that the students had. He found, similarly to Walker and King, that the maps of the expert were different in kind to those of the novices. The study also confirmed that knowledge structures require time and experience to develop. Students cannot be hassled. Further, it supports the contention that teaching should be governed by an understanding of student learning. Teachers should avoid imposing their structures on beginning students. In so far as first year students are concerned, Fordyce felt that "it would be reasonable to expect only simple first level models in relation to a confident 'unified scientific outcome' where it exists." Fordyce draws attention to the need to help students develop confidence, once more underlining the importance of the affective domain, and the context in which real life work takes place [36, 37].

Studies of novices and experts continue to be made. The University of Leeds (UK) and Arizona State University undertook a comparative study of product design and engineering student teams. It was found that a key characteristic of the product design teams were use of drawings throughout the process; in contrast the freshmen engineering teams carried out more detailed information gathering activities. "These differences between senior product and freshmen engineering teams reflected the emphasis in areas of the curriculum" [38]. The study is a reminder of the importance of studies of engineers at work. It also has a bearing on the teaching of engineering in schools (K-12), and the role that visualization has in problem solving [39].

The discussion continues in Journey 6.

NOTES AND REFERENCES

[1] Educational philosophers have debated this at length. Ennis, for example believed in "infusion," Students can be helped by teaching "in which general principles of critical thinking dispositions and abilities are made explicit." McPeck took the immersion view. 57

I have assumed that the arguments for and against teaching critical thinking apply equally to problem solving.

  (i) Norris, S. P. (Ed.), *The Generalizability of Critical Thinking. Multiple Perspectives on an Educational Ideal.* New York, Teachers College Press.

  (ii) Ennis, R. H. (1991). *Critical Thinking.* Englewood Cliffs, NJ, Prentice Hall.

  (iii) McPeck, J. (1981). *Critical Thinking in Education.* New York, St Martin's Press.

[2] Gagné, R. M. (1970). *The Conditions of Learning.* 2nd ed., New York, Holt Rinehart and Winston. 57

[3] Feuerstein, D., Rand, Y., Hoffman, M., and R. Miller (1980). *Instrumental Enrichment.* Baltimore, University Park Press. 57, 60

[4] Lipman, M., Sharp, A. M., and F. S. Oscanyan (1980). *Philosophy in the Classroom.* Philadelphia, Temple University Press. 57

[5] Smith, K. and R. Korte (2008). What do we know? How do we know it? An idiosyncratic reader's guide to philosophies of education. *ASEE/IEEE Proceedings Frontiers in Education Conference*, 1, S4H, 25–28. 57

[6] Grimson, W. (2007). The philosophical nature of engineering. A classification of engineering using the language and activities of philosophy. *Proceedings Annual Conference of the American Society for Engineering Education.* Paper 1611, 14 pages. 57

[7] Red, W. E. (1981). Problem solving and beginning engineering students. *Engineering Education*, 72(2), pp. 167–170. 57

[8] Woods, D. R. et al. (1997). Developing problem solving skills: The McMaster problem solving program. *Journal of Engineering Education* 86(2), pp. 75–91. 57, 58, 61

[9] Heywood, J. (2005). *Engineering Education. Research and Development in Curriculum and Instruction.* Chapter 9, Hoboken, NJ, IEEE/Wiley. 57

See also, for example, Vidic, A. D. (2008). Development of transferable skills within an engineering science context using problem based learning. *International Journal of Engineering Education*, 24(6), pp. 1671–1677.

[10] *ibid* Heywood J. (2005). 57

[11] (a) Jonassen, D. H. (2014). Engineers as problem solvers in A. Johri and B. M. Olds (Eds.), *Cambridge Handbook of Engineering Education Research.* New York, Cambridge University Press. 57, 59
(b) Jonassen, D., Cho, Y-H., and C. Wexler (2017). Facilitating problem transfer in physics. *Proceedings Annual Conference of the American Society for Engineering Education.* Paper 184.

[12] 1984 issue of *Educational Leadership* carries a series of articles related to the topic. 58

[13] Cromwell, L. S. (1986). *Teaching Critical Thinking in the Arts and the Humanities.* Milwaukee, WI, Alverno Productions. 58

[14] McDonald, F. J. (1968). *Educational Psychology.* Belmont, CA, Wadsworth. 59

[15] Sutinen, A. and J. Tarhio (2001). Teaching to identify problems in a creative way. *ASEE/IEEE Frontiers in Education Conference*, 1, T1D-8 to 13. 59

[16] Saupé, J. (1961). Learning in P. Dressel (Ed.), *Evaluation in Higher Education.* Boston, Houghton and Mifflin. 59

[17] Carter, G., Heywood, J., and D. T. Kelly (1986). *A Case Study in Curriculum Assessment. GCE Engineering Science (Advanced).* Manchester, Roundthorn Press. 59, 69

[18] Ruskin, A. M. (1967). Engineering problems for an introductory materials course. *Engineering Education*, 58(3), pp. 220–222. 59, 60

[19] Koen, W. V. (2003). *Discussion of the Method. Conducting the Engineer's Approach to Problem Solving.* New York, Oxford University Press. 59

[20] Koen, W. V. (2010). Quo vadis humans? Engineering the survival of the human species in I. van de Poel and D. E. Goldberg, Eds., *Philosophy and Engineering. An Emerging Agenda.* Dordrecht, NL, Springer. 59, 61

[21] Fordyce, D. (1992). The nature of student learning in engineering education. *International Journal of Technology and Design Education*, 293, pp. 22–40. 60, 69

[22] Kahney, H. (1986) *Problem Solving. A Common Sense Approach.* Milton Keynes, Open University Press. 60

[23] Bolton, J. and S. Ross (1997). Developing physics students, problem solving skills. *Physics Education*, 32(3), pp. 176–185. 60

[24] Thompson, N. (1987). *Thinking Like a Physicist.* Bristol, IOP Publishers. 60

[25] Dean, R. H. and H. L. Plants (1978). Divide and conquer or how to use the problem solving taxonomy to improve the teaching of problem solving. *ASEE/IEEE Proceedings Frontiers in Education Conference*, pp. 268–274. 60

[26] Apple, D. K., Nygren, K. P., Williams, M. W., and D. M. Litynski (2012). Distinguishing and elevating levels of learning in engineering and technology instruction. *ASEE/IEEE Proceedings Frontiers in Education Conference*, 1, T4H-7 to 11. 61

[27] Prince, M and B. Hoyt (2002). Helping students make the transition from novice to expert problem solvers. *ASEE/IEEE Proceedings Frontiers in Education Conference*, 2, F2A-7 to 11. 61

[28] Strobel, J. and M. Cardella (2017). Compound problem solving: Workplace lessons for engineering. *Proceedings Annual Conference of the American Society for Engineering Education.* Paper 450. 62

[29] In Ireland all students of circa 15 years sat a public examination called the Intermediate Certificate. They would take a range of subjects (e.g., Irish, English, Maths, History, etc.), and each subject would set a written paper of between 2 and 2 hours duration. The examinations were set by the Department of Education and controlled by the inspectorate. 61

[30] Heywood, J., McGuinness, S., and D. E. Murphy (1980). *Final Report of the Public Examinations Evaluation Project.* Dublin, University of Dublin, School of Education. 62

[31] *loc. cit.* note [17]. 62

[32] Heywood, J. (2016). *The Assessment of Learning in Engineering Education. Practice and Policy.* Chapter 3, Hoboken, NJ, IEEE/Wiley. 62

[33] Liebold, B. G., Moreland, J. L. C., Ross, D. C., and J. A. Butko (1976). Problem solving. *Engineering Education* 70(3), pp. 285–288. 64

[34] Walker, J. M. T. and P. H. King (2003). Concept mapping as a form of student assessment and instruction in the domain of bioengineering. *Journal of Engineering Education*, 92(2), pp. 167–180. 65

[35] *loc. cit.* note [21]. 66

[36] Atman, C. J. and I. Nair (1996). Engineering in context. An empirical study of freshmen students' conceptual frameworks. *Journal of Engineering Education*, 85(4), pp. 317–326. (See also Borgford-Parnell, J., Diebel, K., and C. J. Atman (2014). Engineering design teams: Considering the forests and the trees. In Williams, B., Figueiredo, J., and J. Trevelyan (Eds.), *Engineering Practice in a Global Context. Understanding the Technical and the Social.* London, CRC Press/Taylor and Francis. 66

[37] Itahashi-Campbell, R. and J. Gluesing (2014). Engineering problem solving in social contexts. "Collective wisdom" and "ba" in Williams, B., Figueiredo, J., and J. Trevelyan (Eds.), *Engineering Practice in a Global Context. Understanding the Technical and the Social.* London, CRC Press/Taylor and Francis. 66

[38] Purzer, S. Y. et al. (2017). Comparing the design problem solving processes of product design and student teams in the U.S. and the UK. *Proceedings Annual Conference of the American Society for Engineering Education.* Paper 1709. 66

[39] Uria, E. S. and M. G. Mujika (2017). Teaching part visualization in first year engineering courses: Methodology for part visualization problem solving. *Proceedings Annual Conference of the American Society for Engineering Education.* Paper 124. 66

JOURNEY 6

Critical Thinking, Decision Making, and Problem Solving

6.1 INTRODUCTION

Leibold and his colleague students expressed the need for "hints" and "heuristics" to help them with problem solving (Journey 5). Perhaps the best known heuristic is due to Polya [1]. In engineering it was adapted by Red [2], and Rosati used it to design computer routines for problems in engineering statics [3]. A modified version was also used in the McMaster Problem Solving Course in engineering [4]. Fuller and Kardos explain how maps, known as Polya maps, can be used to define problems in engineering [5]. However, in 1986 Wales and Stager published a series of articles in *Engineering Education* on Guided Design [6], a decision making heuristic that they regarded as universal, that is, it may be applied within any subject of the curriculum (Exhibit 6.1).

Wales and Stager's Guided Design Heuristic	Polya's Problem Solving Heuristic
1. Define the situation	1. Understanding the problem
2. State the goal	2. Devising a plan
3. Generate ideas	3. Carrying out the plan
4. Prepare a plan	4. Looking back
5. Take action	
(6. Look back)	

Exhibit 6.1: The Wales and Stager and Polya heuristics. Item 6 shows that many students in the reported study modified the Wales and Stager heuristic.

Wales and Stager argued that because guided design is part system and part attitude it is important to pay attention to the needs of the students. In their model these needs are perceived to be hierarchical, and ordered as in Maslow's model of motivation [7]. The course "is based on the conviction that the student who works through an ascending order of well-designed problems, who is actively seeking solutions to problems rather than passively assimilating knowledge, will emerge not only better educated but far stronger intellectually" [8]. The problems are chosen to be relevant and interdisciplinary.

During their decision making course the students "work in small groups (and) attack open ended problems rather than masses of information." The course is structured so that each problem creates the need for subject matter that has to be learned independently by the student out of class time. Its purpose is to show that knowledge of concepts, principles, and values is necessary. The teacher is a facilitator who in part listens and encourages the students to participate in the decision making process, and in part by asking them leading questions.

From their Center for Guided Design they published books to support their case [9]. They also published a series of Sherlock Holmes stories designed to illustrate the heuristic in action. In these situations the decision maker (the detective) had to ask, when defining the situation, who is involved? (The actors). What things are involved? (The props). What happened? (Cause). How serious is its effect? (Effect). These are questions that help the decision maker "learn" about the situation, and in this sense this decision making model is also a model of learning. Thus, we might conclude, that if students use such a model of decision making, they are likely to enhance their learning.

6.2   TEACHING A DECISION MAKING HEURISTIC

With this in mind, and since they claimed that the model is generalizable, I asked all my graduate trainee teachers to evaluate a model of decision making with the strong suggestion that they should consider the Wales and Stager model [10]. Most did, but some preferred the Polya heuristic. Some modified the Guided Design model by adding a 6th stage, for example, "Look back (was your decision a good one? Is there a better answer)."

In the year that their work was evaluated in detail, it was not possible to say that the students' decision making skills were improved as a result of the exercise, but there was evidence of an improvement in average test performance among average and weaker students. Most of the student teachers (80%) thought the exercise had been of value in developing decision making skills, but just under half thought that only some of their school students greatly benefited.

In previous years the student teachers had reported that low achieving pupils benefited the most from the exercise. They thought that it helped them retain concepts, and that their self-esteem was enhanced. One reason for this might be that the weak pupils benefit from the structure imposed on their study by the heuristic, although not every pupil liked the planning that it required. For these pupils it might also induce "set mechanization;" (see below) but, it can be argued that it is better for the weaker student to solve problems rather than to give up. The examples show that the heuristic was a powerful diagnostic tool since it revealed that weaker pupils find it difficult to formulate a problem, or distinguish between relevant and irrelevant information. This may be found to be true of pupils functioning at much higher levels of abstraction. Some of these brighter pupils did not see the value of the heuristic. It seemed that it clashed with their own ways of thinking. It seems to me that it would safe to substitute undergraduate for pupil in the above, certainly freshmen.

These exercises confirmed that the heuristic may be applied in any subject of the curriculum (Exhibit 6.2). In this example the notes for stage 4 indicate that the teacher had made use of the literature that the class had been given about other decision making models. Question 1 of the test however, suggests that the pupils might be given to understand that there is only one way of solving problems. This seems to be the case with question 1 of the test shown in Exhibit 6.3. Students need to be shown a variety of ways of solving problems otherwise they can easily fall into the trap of "set mechanization" in which the pupils always use the same model of problem solving even when it is inappropriate [11].

One student teacher of business studies to 16 year olds did recognize this problem. She argued that some problems would not be best solved by the Wales and Stager approach, and she gave an example of a problem where there could only be one right answer for which a compensatory approach would be more appropriate. She went on to say that "the importance of set is that the teacher should try to design problems which have several different methods or solutions so that the pupil becomes aware that the most complex problems, and indeed the most simple ones, can be solved in different ways. If they become familiar with this when facing new problems they will realize there may be more than one way to solve it."

Just how difficult this may be, is illustrated by another teacher of business studies to the same age group. She reported that even though the pupils responded positively to the teaching of the decision making heuristic, and were able to recall various stages, when it came to actually attempting the second game, the heuristic was abandoned... "They didn't use the systematic approach reverting instead to their basic instincts." As Peter Lydon (whose report has been published in full elsewhere) put it "the knowledge they acquired is by no means permanently in their heads. Only by continued exposure to this type of exercise, or better still, this method of teaching will the pupils be able to become critical thinkers, or at the very least effective decision makers." To achieve that goal teachers' would have to plan a curriculum not just a lesson [12].

In this study the use of the heuristic changed the role of the teachers as they responded to active learning, and for some it provided a structure for their learning, and others the beginning of meta-cognition.

6.3   QUALITATIVE STRATEGIES

Studies of experts and novices (e.g., Larkin [13], and Journey 5) showed that there was something more to problem solving than the learning of a range of heuristics [14]. It was suggested that a major learning impediment to solving engineering problems is the inability of undergraduates to use qualitative strategies in problem solving.

Cowan [15] found that engineering students handled qualitative analysis ineffectively. He used protocols to help him understand what was happening. It led him to develop a style of tutorial question which "literally demanded qualitative understanding and offered no return for

The Heuristic-Outline of Instructions (handout provided)	The Test
1. Define the situation Given a set of poems of varying lengths of which one could be a Shakespearean sonnet. What features are involved? List the features	
2. State the goal What do I have to do? i.e. find out if these poems are Shakespearean sonnets.	
3. Generate ideas Consider different ways you can achieve your goal. Will you count the lines? Will you go by technical characteristics of a Shakespearean sonnet? Are they all important?	
4. Prepare a plan How are you going to decide? Suggestions: consider the features which a Shakespearean sonnet must have. List them. Match your two poems against them Give +3 if totally agrees, and -3 if totally disagrees with the feature. Compromise with values between +3 and -3 if it half agrees or disagrees. Add up the scores and see which is the greatest value. Poem with the greatest value is a sonnet. Or, use a process of elimination by aspects. Have the minimum criteria which the poem must have in order to be a Shakespearean sonnet been met? Put these attributes in order of importance. Check attributes against each possibility and eliminate where the poem does not meet the criteria.	**The Test** 1. List the 5 steps we can take when we want to solve the "problem" of approaching a poetry question (20 marks). 2. Briefly discuss what each step involves (20 marks). 3. Using the 5 step strategy and illustrating reasons for your choices, write your approach to the following question: What is the salient theme of Dickinson's poem *Because I could not stop for death*? (60 marks). (the pupils were provided with a copy of the poem)
5. Take action Make your decision. Is this a Shakespearean sonnet?	

Exhibit 6.2: A graduate student teacher's attempt to apply the Wales and Stager heuristic to the teaching of English to a group of 15–16 year olds.

Decision Making Test

1. When we want to solve a problem, there are six steps we can take. What are these steps and what do they involve?

2. Look at the map provided. Imagine you are a planning officer and you have to decide where to locate an industrial estate. Write down five questions you should ask yourself in trying to decide on a location.

3. You have now chosen your locations, however you can only build one estate. Fill out the model below and choose one location.

4. Please write your reasons for choosing that location and say whether or not you are happy with it. Would you like to change your mind?

5. Please say whether you like or disliked being taught decision making in Geography. Don't be afraid to say you if you disliked if that is the case?

Exhibit 6.3: A test that was set to 12–13 year olds after a lesson on location in geography [12].

quantitative understanding in other words, I introduced problems where a solution could not be obtained merely by applying formulae and carrying out calculations, but called instead for the application of deep conceptual understanding." He went on to say, "These problems incidentally, often proved insuperable for conventional lecturers accustomed to following algorithms rather than thinking. Quantitative understanding that is the result of routine calculations does not necessarily require a sound grasp of the concept whereas qualitative understanding does (Exhibit 6.4), which is why understanding how novices learn concepts is important" (see Journey 5). Note [20] of Journey 4 gave details of one of the assessment questions he used to test qualitative thinking.

In the United States McCracken and Newstetter [16] drew attention to another difficulty in thinking both qualitatively and quantitatively that arises from the fact that engineers have to speak a number of "languages." They propose the use of "representational transformation." Such transformations are "built upon community-sanctioned practices often referred to as 'back of envelope' calculations." Engineers often do such calculations. They begin with a problem statement, which is then transformed into a problem as they relate to each other; it is then finally translated into a set of mathematical formulae. The diagrammatic account is qualitative.

McCracken and Newstetter point out that the knowledge necessary to undertake these representational transformations is central to engineering practice. They argued that one of the reasons why students may find problem solving difficult is that since each representation uses different symbols, and is therefore a different linguistic system, engineering students are faced with having to learn three different languages. "Multiple literacies are required to do engineering problem solving." This is evidently related to the solution of compound problems discussed by Strobel and Cardella in Journey 5.

"You will recall that I asked one boy at this school to try out the diode valve experiment of the semi-problem type that we discussed and which your read. As a matter of interest, I asked him today to design an experiment to verify a hypothesis (which he was to formulate) regarding water discharge through an orifice in terms of pressure difference or total pressure. He has not met this topic before.

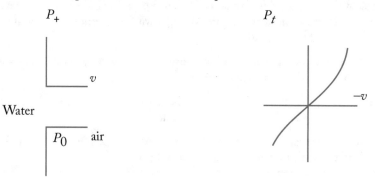

The result surprised me! He suggested that the velocity of discharge would increase more rapidly than the P.D. Why? Because "it seems reasonable." What form of increase? It is "that the velocity of trans-ference from one pressure to another lower pressure is proportional to the square of the pressure difference"—again "because it seems reasonable." Note that the mathematical formulation is wrong in terms of the physics. His graph agrees with the maths, not with the physics and includes both positive and negative values of flow."

Exhibit 6.4: A letter about the thinking of a student of engineering science sent to this writer by Glyn Price that illustrates the problem of qualitative thinking highlighted by John Cowan.

6.4   CRITICAL THINKING

Nowhere are differences between educational cultures more exposed than in considerations of critical thinking. In the United States, critical thinking is often assessed by using standardized tests [17]. In the British Isles many academics would say, "Well our tests (teacher designed) do that anyway." Critical thinking is acquired by osmosis. Cajander and his colleagues at Uppsala University in Sweden write, "it is not uncommon to view competencies such as critical thinking and communication as something that develops as a side effect while learning is the knowledge associated with a subject e.g., computer science." For which reason, much of the literature that relates critical thinking to assessment is from the United States.

As with assessment, the concept suffers from the fact that not all those who study critical thinking agree about what it means [18, 19]. The focus of some participants is on inductive and deductive logic, whereas others simply consider it to be a process of problem solving. As might be expected, this is the perspective that has been adopted by some engineering educators.

Pascarella and Terenzini in their mighty study of the effects of college write, "it would appear that most attempts to define and measure critical thinking operationally focus on an individual's capability to do some or all of the following: identify issues and assumptions in an argument, recognize important relationships, make correct references from the data, deduce conclusions from the information or data provided, interpret whether conclusions are warranted based on given data, evaluate evidence or authority, make self-corrections, and solve problems" [20, p. 156]. The similarities with discussions about problem solving and decision making will be apparent.

The Queens University, Canada, in a striking change to its first year engineering curriculum integrated Model Eliciting Activities (MEA) into the course. MEA's seem to have had their origins in mathematics. An MEA involves a case study of 30–50 minutes which is solved by students in groups of between 3 and 5 individuals. MEAs are intended to stimulate real world problems for which testable or model solutions may be found.

The curriculum principle underlying these activities is to begin with the student's own conceptual system as a basis for modeling that is, "creating representations of problematic phenomena or scenarios as means to solve those situations." This would seem to be what a student has to do if he or she is asked to undertake a mini project or investigation of their own creation, and pursue it to its practical conclusion without formal instruction [21].

Be that as it may, an evaluation of the impact of the MEA course on the development of critical thinking skills came to the conclusion that the training given during the course enabled a relative improvement in critical thinking when compared with the results of a control group. The principles of critical thinking had been discussed in class in each MEA. In one class the skill of questioning was developed by creating lists of questions that should be asked by the accident investigators. This is of course a contributory component of the skill of "diagnosis" which is seldom discussed in the engineering literature. However, its significance was made clear by Lin, Shahhosseini and Badar (2017) at the annual conference of the American Society for Engineering Education [22].

A major problem in the modern university is compartmentalisation of subjects. This can mean that subjects that could collaborate in the development of critical thinking either don't, or do not collaborate, as for example in courses where engineering students are required to take subjects within the liberal arts. Yet, the potential for collaboration in the development of critical thinking skills is great as the example in Exhibit 6.5 shows.

Clearly, reflective thought is important in critical thinking, if it is not then a problem is created. But, that is a matter for another discussion.

6.5   A CATEGORY FOR PROBLEM SOLVING?

It is evident that assessment plays an important and integral role in enabling students to acquire problem solving skills, moreover it is also evident that it is difficult to write questions that perform this function [23]. The list of learning objectives shown in Exhibit 6.6 are reminder that

In Your Work	In assessing someone else's work (especially when acting as an adviser, assessor or in a debate)
Outlining the argument List the premises (hypotheses, propositions) which you wish to demonstrate or prove. State the conclusions. List the reasons for the conclusions.	**Outlining the argument** Identify the premises (hypotheses) and conclusions. Establish which sentences do not add to the argument. List the reasons for the conclusion.
Examining the argument for clarity Check the key terms (concepts) principles are stated clearly. Do not attempt to fudge the issue. NB an argument whose key concepts and principles are not clear, or which contains ambiguities is not a good argument. A good argument should be brief and to the point.	**Examining the argument for clarity** When terms and phrases are unclear ask for clarification in a live argument. In an examination write a note at the side of the script to check you have not misperceived what is being said. Look out for fudging.
Asserting and checking the truth of premises You must be able to stand over each premise you use, and where necessary cite supporting evidence and its course. Don't throw false premises into an argument because they are easily spotted	**Asserting and checking the truth of premises** Any questionable premises must be defended by another argument. In an examination answer look for false premises as a foundation for your own assessment. List the reasons which make you think that the proposition is false (see last section below).
Ensuring the premises are necessary and relevant to the argument Check that each premise is both necessary and relevant to the argument. Eliminate premises which are not. NB any person involved in a disputation must have sure grounds for the claims that are made.	**Ensuring that the premises are necessary and relevant to the argument** In debate the person offering the argument must be able to show the necessity of and relevance of each answer. In scripts look out for circular argument and tautologies.
Testing the strength of an argument Predict the counter argument and test your conclusion for validity against the alternative. If the alternative view has merits then your conclusion may be wrong.	**Testing the strength of an argument** In debate and examination scripts look for the weaknesses in the logic of the argument, and the data used to support the conclusion.

Exhibit 6.5: Evaluating arguments in essays and debates. Adapted from R. E. FitzGibbons (1981). *Educational Decisions. An Introduction to the Philosophy of Education*, New York, Harcourt, brace, Yovanovitch.

there are a range of activities that can be used to develop problem solving and critical thinking skills, especially when dealing with practical engineering problems.

Given a term listed under "concepts introduced," you should be able to give a word definition, list pertinent characteristics and cite an example. You will be able to describe d-lines, describe the limitations of short-term memory, and rationalize the processes used in brainstorming.

Given an object or a situation, as an individual you will be able to generate at least 50 uses, attributes, or ideas in 5 minutes.

Given an object or a situation as an individual you will be able to generate at least 50 ideas in ten minutes and the ideas will belong to at least 7 different categories, and a group of three independent shall identify one idea that is "unique."

Given a crazy idea, you will be able to describe your mental processes used to convert that idea into a technically feasible idea by using the triggered idea as a "stepping stone."

You will be able to describe your preferred style of brainstorming and your preferred use of triggers.

Exhibit 6.6: Five of the ten learning objectives for the McMaster Problem Solving Unit on Creativity as they relate to the assessment [4]. (Reproduced with the permission of D. R. Woods).

In Journey 4 the value of, and difficulties associated with *The Taxonomy of Educational Objectives* were considered. One of these, given the importance attached to problem solving by engineers, is the lack of a specific category for problem solving. Journey 5 and this journey followed this up with the purpose of examining whether or not there is the possibility of a specific category of problem solving. It is shown that there is a considerable focus in engineering education on problem solving, and it is submitted that the case for a distinct category of problem solving in engineering education is self-evident from the information provided.

6.6   LOOKING BACK OVER JOURNEYS 4, 5, AND 6

In Journey 4 the social efficiency ideology of the curriculum was introduced for the reason that it prevails in many industrialized nations. It has its origins in the objectives (outcomes) movement which is supported by educators who believe that knowledge is defined behaviorally in terms of what students will be able to do as illustrated in *The Taxonomy of Educational Objectives*. While some engineering educators use *The Taxonomy*, others believe that it, and its revision do not deal adequately with problem solving, decision making and critical thinking. They would like to see a distinct category of problem solving.

Journeys 5 and 6 explore developments in problem solving in engineering education. It is concluded that a substantial case may be made for such a category.

In reviewing the engineering literature on the subject, a number of issues/questions emerged. The first related to the impact of assessment on learning, and in particular the design of questions to test higher order problem solving. Second, the recognition that changing the conditions of learning impacts on the role of the teacher. Third, there are questions about who an instructor's students are? What should instructors know about their students? Fourth, there are questions about the "time required" to achieve certain objectives. The fifth issue related to the impact of the structure of the organization both institutionally and at the classroom level on learning: and the sixth related to the importance of qualitative thinking and that students find it difficult to think qualitatively. Finally, the impact of teacher beliefs on what they do was noted.

For many teachers these beliefs may be described as belonging to a scholar-academic ideology. But some of those seeking to improve problem solving have promoted specialist courses that are problem based. These courses clearly belong to a different ideology, and a different understanding of "knowledge." Nevertheless, they do not wholly escape from the Scholar Academic Ideology or the need to ensure that key concepts are understood. In the next journey we will consider the Scholar Academic Ideology and its impact on the curriculum.

NOTES AND REFERENCES

[1] Polya, G. (1957). *How to Solve it*, 2nd ed., Garden City, Doubleday Anchor. 71

[2] Red, W. E. (1981). Problem solving and beginning engineering students. *Engineering Education*, 72(2), pp. 167–170. 71

Red used this approach in laboratory work. He said that the model does not work unless the students are given detailed instructions. Thus, at the plan stage the students were instructed to write down the equations from which the unknowns can be found using the given information. These equations are the connection between the given data and the unknowns. Then

(1) Carry out the plan, i.e., apply the equations.

(2) Check each step.

(3) Underline or block in each answer so that it is easily identified.

(4) Make sure that each answer meets with common sense, i.e., is it realistic? Then make sure each answer satisfies the assumptions and conditions stated for the problem.

[3] Rosati, P. A. (1987). Practising a problem solving strategy with computer tutorials. *International Journal of Applied Engineering Education*, 3(1), pp. 49–53. 71

[4] Woods, D. R. et al. (1997). Developing problem solving skills: The McMaster problem solving program. *Journal of Engineering Education*, 86(2), pp. 75–91. 71, 79

[5] Fuller, M. and G. Kardos (1980). Structure and process in problem solving in J. Lubkin (Ed.), *The Teaching of Problem Solving in Engineering and Related Fields*. Washington, DC, ASEE. 71

[6] Wales, C. E. and R. A. Stager (1986). Series of papers in issues, 5, 6, 7, and 8 in volume 62 in *Engineering Education*. 71

It should be noted that there are other approaches that use a design approach for problem solving and decision making, as for example the "backward design process" and, "reverse engineering." See for example, Sgro, S. and S. Freeman (2017). Teaching critical thinking using understanding by design. *Proceedings Annual Conference of the American Society for Engineering Education*. Paper 924. The process begins with identifying the results, acceptable evidence is then determined, and finally the learning experiences are planned.

Typical examples of this are the reports that are held into the reasons for major disasters. In the UK at the present time (July 2017) there are public criticisms of the terms of a major enquiry to be conducted into a fire in a high rise block of flats that killed nearly 100 individuals. There are also criticisms of the way investigations into the materials used for cladding (thought to be the cause of the fire) are being conducted.

[7] Maslow (i) makes a distinction between primary and secondary needs. The primary needs are basic, such as water, food, sex, lactation, urination, defecation, heat avoidance and cold avoidance. The secondary needs vary from person to person. They are ranked hierarchically in the sense that the primary needs have to be met before the secondary needs can be met. 71

There are five levels in the hierarchy the first of which are the physiological needs listed above, these are followed by safety needs, love and belonging needs, esteem needs, and self-actualisation needs. The teachers task is to arrange that the student works in an environment (with appropriate resources) that will help the individual strive toward self-actualisation (i.e., become everything that one is capable of becoming). Barnes, does not believe that self-actualisation is operationally possible in industry. He also argues that safety needs overlap the higher needs because whenever one of these is threatened, so is the individual's safety (ii).

(i) Maslow, A. (1964). *Motivation and Personality*. New York. Harper and Row.

(ii) Barnes, L. B. (1960). *Organizational Systems and Engineering Groups*. Cambridge, MA, Harvard Business School.

[8] D'Amour, G. and C. A. Wales (1977). Improving problem solving skills through a course in guided design. *Engineering Education*, February, pp. 381–384. Wales, C. E. and R. A. Stager (1986) issues 5, 6, 7, and 8 of Vol 62 of *Engineering Education*, contain a series of articles on guided design. 72

[9] Wales, C. E., Nardi, A. H., and R. A. Stager (1986). *Professional Decision Making*. Morganstown, WV, Center for Guided Design, West Virginia University. 72

[10] Heywood, L. (2008). *Instructional and Curriculum Leadership. Towards Inquiry Oriented Schools*. Dublin. Original Writing for the National Association of Principals and Deputies. See also Heywood, J. (1996). An approach to teaching decision making skills using an engineering heuristic. *ASEE/IEEE Proceedings Frontiers in Education Conference*, 1, pp. 67–73. 72, 82

[11] Luchins asked pupils in his research group to obtain specified amounts of water from three jars filled to different levels (capacity). He showed them first two problems which involved all three jars. In a further nine problems, the subjects mainly used the three-jar solution, to solve the problem when two jar solutions were possible. The "set" interfered with their problem solving. Subsequently Luchins divided another group into two sub-groups. The first sub-group worked through the problem in the usual way, using the three jar solution. The second sub-group were told to think more carefully about how to solve the problems. Given that instruction, the majority of students in the second sub-group moved to the more simple solution using two jars. 73

Set mechanization is also called set induction.

Adapted from a description of Luchins study by F. J. MacDonald (1968). *Educational Psychology*. Belmont. CA, Wadsworth.

Luchins, A. S. (1942). Mechanization in problems solving: The effect of "Einstellung." *Psychological Monographs*, no 248.

[12] The quotations in this paragraph come from Heywood (2008) [10]. More details of Lydon's plan, which was made available to future classes will also be found in that reference. 73, 75

[13] Larkin, J. H. (1979). Processing information for effective problem solving. *Engineering Education*, 70(3), pp. 285–288. 73

[14] Marton, F., Hounsell, D., and N. J. Entwistle (Eds.), (1984). *The Experience of Learning*. Edinburgh. Scottish Academic Press. 73

[15] Cowan, J. (1998). *One Becoming an Innovative University Teacher. Reflection in Action*. Buckingham. SRHE/Open University Press. 73

[16] McCracken, W. M. and W. C. Newstetter (2001). Text to diagram to symbol. Representational transformations in problem solving. *ASEE/IEEE Proceedings Frontiers in Education Conference*, F2G-13 to 17. 75

[17] In the United States, among the most commonly used tests of critical thinking are the California Critical thinking Skills test, The Watson-Glaser Critical Thinking Appraisal, and the Cornell Critical Thinking test. All three are composed of multiple choice items,

but there are issues of validity associated with each test (i). In the engineering literature Kaupp,and Frank have provided detailed descriptions and critiques of the Cornell test, the Paul Elder Model, the CLA model (Collegiate Learning Assessment) (ii). Stein collaborated with faculty to develop a test that was more directly related to faculty objectives, and so engage faculty in quality enhancement. Unusually the test used primarily essay questions that would enable the assessment of communication and "leave opportunities for creative answers to questions that don't always have a single correct response." Faculty were trained to rate the test answers which were each graded by two raters (i). 76

A Portuguese study examined the ability to argue. It found among first year engineering students a preference for abductive (inference to the best explanation) and deductive reasoning. It found that the argumentative characteristics are dependent on the methods of assessment which may need to be changed. But the authors considered their results reflected the lower level of maturity of the students. They, suggested that the ability to argue is developed progressively, and that course structures have to be designed to meet this need (iii).

(i) Stein, B. and A. Haynes (2011). Engaging faculty in the assessment and improvement of students' critical thinking assessment test. *Change,* March, April, 2011.

(ii) Kaupp, J. A. and B. M. Frank (2014). Potential, authentic and sustainable development of critical thinking in education through model eliciting activities. *Proceedings Annual Conference of the American Society for Engineering Education.* Paper 9806.

(iii) Liete, C. et al. (2011). A place for arguing in engineering education; a study of student's assessments. *European Journal of Engineering Education,* 36(6), pp. 607–616.

[18] Cajander, A., Daniels, M., and B. R. Komsky (2011). Development of professional competencies in engineering education. *ASEE/IEEE Proceedings Frontiers in Education Conference,* S1C-1 to 5. 76

[19] A recent review of critical thinking will be found in, Cooney, E., Alfrey, H., and S. Owens (2017). Critical thinking in engineering and technology. A review. *Proceedings Annual Conference of the American Society for Engineering Education.* Paper 1110. 76

[20] Pascarella, E. T. and P. T. Terenzini (2005). *How College Affects Students. A Third Decade,* vol 2, San Fransisco. Josey Bass. 77

[21] Kaupp, J. A., Frank, B. M., and A. S.-Y. Chen (2013). Investigating the impact of model eliciting activities on the development of critical thinking. *Proceedings of Annual Conference of the American Society for Engineering Education.* Paper 6432. 77

[22] Lin, Y., Shahhosseini, A. M., and M. A. Badar (2017). Assessing conceptual mapping based active learning for advancing engineering diagnostic skills. *Proceedings Annual Conference of the American Society for Engineering Education.* Session T149. 77

[23] The Joint Matriculation Board published each year reports of how the candidates performed in each of the questions. One comment on a question in engineering science was published by Carter, Heywood and Kelly (i). It illustrates just how difficult it is to design such questions. The report reads: "The figure (not shown here) represents some of the more important parts of a single bar, 1 kw radiant electric fire. Discuss the purpose of these components and suggest a suitable material for each. (Base your discussion on the function each part has to fulfil and requisite physical properties). Discuss the other factors that a manufacturer would consider in producing the components from particular materials. Describe and suggest materials for other parts which you believe will be necessary for satisfactory use of the fire, but which has not been indicated in the sketch." 77

The examiners published comment on the answers was: "This was the most popular question and the most badly done. Only one of the two candidates calculated the resistance required for the element. Candidates tended not to answer the questions asked. E.g., they did not state the function of each of the parts of the fire and materials were often suggested without reasons. Candidates stated factors the manufacturer should consider, without discussion. The question was answered on the whole in too facile a number."

In (i) they expanded on this type of question as follows, "Although this type of question suffered from superficiality of response it was retained in similar form as a component of the examination for six years. However, significant attempts were made to direct candidates' answers into more detailed engineering analyses of the problems set, by the requiring statements relevant, for example, to improved safety and efficiency, broadening the range of use or versatility of the device, and by specifying more closely, the parameters which were of most importance, e.g., electrical, mechanical, thermal or optical properties. Although this further guidance was given, the Examiners' reports continued to indicate that a significant proportion of the candidates' answers were superficial, and that the necessary skills for attempting such questions were not being fully developed by the curriculum study as hoped. In the seventh year, therefore, this type of question was modified to consider not engineering devices, but engineering situations, and the methods of achieving solutions under a variety of constraints. Thus, in the next examination a question was set about the design of a technician's preparation room and the modifications to the design which would be necessary by imposition of a 50% reduction in available finance after the first design stage. The topic was deliberately chosen to lie within the familiarity and experience of the candidates; logical argument and judgement about possible alternative solutions were required from the candidate. The realities of life were introduced into the question through economic constraints and the skills of evaluation and judgement were tested. This type of question is generally, most difficult to assess but with experience the examiners are readily able to evaluate the cogent and relevant arguments and detect the simplistic and facile. Since the introduction of this type of question, there have been many excellent answers and there have been some signs of a general improvement in the candidates' engineering reasoning, synthesis and evaluative ability."

(i) Carter, G., Heywood, J., and D. T. Kelly (1986). *Case Study in Curriculum Assessment. GCE Engineering Science (Advanced)*. Manchester. Roundthorn Press.

JOURNEY 7

The Scholar Academic Ideology of the Disciplines

7.1 INTRODUCTION

We all have beliefs, some unconsciously held, about how children and students learn, and what the curriculum should be. These beliefs are, to some extent or another, conditioned by the culture in which we live, and the education it promotes. They dictate the attitudes we take toward the curriculum, instruction and learning.

While it is of immense value to understand how other systems work, not least from the need to communicate in a globalized world, it is extremely difficult to translate what appears to be effective in one system to another. At the same time one or two educational ideas have pervaded the industrialized nations, one of which is the idea of educating students to achieve specified outcomes. It fits the ideas of business people, and in consequence is attractive to politicians who wish to get value for money from the education systems they finance. Michael Schiro, an American educator has suggested that those who subscribe to this belief possess a "social efficiency ideology" (see Journey 4) [1].

Tensions may be created between teachers and administrators who hold different educational ideologies which may range across a spectrum from those considered to be traditional to those considered to be progressive, although neither term is seldom, if at all, adequately defined.

7.2 THE RECEIVED CURRICULUM OR THE SCHOLAR ACADEMIC IDEOLOGY

John Eggleston, an educational sociologist and technical educator, has described three paradigms that help us to understand the issues involved in the possession of these ideologies. The first of these he calls "received" [2]. In this paradigm of the curriculum knowledge is received, and accepted as given. It is non-negotiable, non-dialectic, and co-sensual. Knowledge is something that is given, and is that which should be transmitted to children and students. Through it the accumulated wisdom of a culture is transmitted.

The "received" view of knowledge has its origins in ancient philosophy (essences and ideal forms) [3, (i)], and in the notion of fixed structures of thought as suggested in the educational psychology of Jean Piaget [4]. Scholar academics are faced with the problem of classification, which was also an issue faced by the Greek philosophers. It asks the questions, "What constitutes a discipline?" And, "what is the relationship between the disciplines?" [3, (ii)]. This question is pertinent to engineering for Rosalind Williams has asked, "Is there a discipline of engineering?" [31]. "If so what are the relationships of areas of knowledge like bio-engineering to engineering?"

The legislated national curriculum for schools in England and Wales reflects this ideology: It is closely associated with what we understand to be the academic disciplines. The national curriculum reflects the importance society ascribes to certain disciplines, and their particular ways of knowing. Schools are allowed to provide other studies for a limited time during each week. Some would argue that, taken together, the disciplines of the national curriculum provide the information necessary to survive (some would say thrive) in the modern world [5].

Whitfield argues that what is good, and what is worthwhile, are learned within the disciplines [6]. Truth is contained within the disciplines, so if students are to learn the truth, they have to be initiated into the disciplines [7]. Each discipline seeks to mould students in its own image and likeness. Subject matter is the essence of the disciplines.

Michael Schiro calls this set of beliefs the "Scholar Academic ideology." "Scholar academics assume that the academic disciplines, the world of the intellect, and the world of knowledge are loosely equivalent. The central task of education is taken to be the extension of the components of this equivalence. Both on the cultural level, as reflected in the discovery of new truth, and on the individual level, as reflected in the enculturation of individuals into civilization's accumulated knowledge and ways of knowing" [8].

Jerome Bruner a distinguished American psychologist wrote: "A body of knowledge enshrined in a university faculty and embodied in a series of authoritative volumes is the result of much prior intellectual activity. To instruct someone in these disciplines is not a matter of getting him to commit results to mind. Rather it is to teach him to participate in the process that makes possible the establishment of knowledge. We teach a subject not to produce little living libraries on that subject, but rather to get a student to think mathematically for himself, to consider matters as a historian does, to take part in the process of knowledge-getting. Knowing is a process, not a product" [9].

The process that makes possible the establishment of knowledge is, in this ideology, what is understood by learning. For each school subject there must be a corresponding academic discipline. Because the disciplines are dynamic they are concerned as much with "what will be?" as with "what was?" [10]. This point is illustrated by the great curriculum projects that were undertaken in the sixties and seventies. These were initiated because teachers did not have the resources to undertake such developments; yet, in the scholar academic view such developments are considered to be part of the role of the teacher [11].

The scholar academic ideology is teacher centered. Information is conveyed to the mind which reasons about it as required. Learning is the result of teaching [12]. Because each discipline has within it, its own theory of learning, generalized theories of learning have no place in the design of instruction. It is this view that is challenged in this book, and it is this controversy that beginning engineering educators should be asked to resolve.

7.3   THE POST SPUTNIK REFORM PROJECTS

The launch of the first two artificial earth satellites known as Sputniks by the Soviets in 1957 caused consternation in the western world. It resulted in a review of school education in the United States because it was believed that American education was falling behind the rest of the world. William Schubert comments that the post Sputnik curriculum projects pushed "the curriculum balance toward the disciplines of knowledge in the interest of social and political ends" [13]. Jerome Bruner played a major role in these developments which led in 1959, to the Woods Hole Conference sponsored by the National Academy of Sciences. His account of the proceedings became a best seller [14].

In this book he enunciated three principles of learning:

1. Children will learn more effectively if they discover the concepts and principles for themselves. Hence, the development of discovery or inquiry based learning.

2. Children can understand the most profound ideas provided that they are presented in a language that is relevant to their development. Other educators with this view included Matthew Lipman who developed the philosophy for young children movement, Gareth Matthews who wrote *Philosophy and the Young Child* [15], and Russell Stannard a Professor of Physics at the Open University who wrote books for young children on modern physics [16].

3. The child should focus on the structure of the disciplines and not on the simple acquisition of information.

Substitute student for child and the questions become relevant to engineering educators. Does inquiry based learning have a role in their course? Why if young children are capable of learning and discussing philosophical issues do so many engineering students remember, but do not learn? It is a complaint that I often hear from engineering educators.

Bruner applied these principles when he designed MACOS (Man a Course of Study) for elementary school children. This curriculum program was intended to introduce students to the disciplines of Anthropology, Ethnography and Social Psychology [18]. The programme was funded by the National Science Foundation between 1963 and 1970.

One criticism of the third principle is that some academic disciplines, it is argued, do not have an inherent structure. Among the subjects cited to illustrate this point are French literature, English poetry, American History, Psychology, Sociology, Modern Painting, and Business Management, a list, which is to say the least, controversial. Schubert writes that, "apart from a few basics, it is even difficult for experts in mathematics and the hard sciences to agree on the dimensions of the structure of their discipline. While some learners may develop an intuitive grasp of a field of inquiry, this is much more elusive that something that can be taught as a matter of fact. The failure of the 1960s curriculum reform projects supports this criticism" [19]. Schubert causes us to ask "what is the structure of the engineering discipline?"

However, the idea of "discovery," more often than not now called "inquiry" persists.

7.4   DISCOVERY (INQUIRY) BASED LEARNING

There was nothing new about discovery or inquiry learning. Socratic questioning is a process of discovery. But, Shulman argues of the learning by discovery approach that emerged during the nineteen-sixties that more than any one person Bruner "managed to capture its spirit, provide it with a theoretical foundation, and disseminate it" [20].

The fundamental principle behind Bruner's approach is that when a person is engaged in the process of discovery, they discover something from within themselves that enables them to reorganize previously known ideas, concepts, and principles if you prefer, and the patterns they encounter in the world. In Bruner's interpretation of the Piagetian stages the child moves sequentially through three modes or representations which he calls: the enactive; the ikonic; and, the symbolic.

In the "enactive" mode learning takes place through action rather than words. Bruner thought that conditioning and stimulus-response learning are appropriate to this mode. In the example that he gives which relates to mathematics, the children play with the materials, in this case three flat pieces of wood [21]. Once they have the feel of the materials they are asked if they can make a larger square than the largest wooden square. They are then asked to describe what they have done, and so on. Bruner hypothesizes that at some point they will perceive a pattern.

In the "ikonic" mode, concrete visual imagery is deployed; the materials are no longer manipulated. In the "symbolic" mode, the learner manipulates symbols. Shulman writes: "The combination of these concepts of manipulation of actual materials as part of a developmental model, and the Socratic notion of learning as internal reorganization into a learning-by-discovery approach is the unique contribution of Bruner" [22].

One reason for engineering educators to take note of Bruner's approach, is the idea of structure in lectures, and of moving toward the abstract from the concrete. Also, it may be argued that these representations are at the heart of the mental process of design.

Inspection of the 2017 proceedings of the Annual Conference of the American Society for Engineering Education suggests that the term "inquiry" has replaced "discovery" [23]. But, it seems to have the same difficulties of definition that Shulman recorded for discovery learning. Describing. Wittrock's classification of discovery learning as shown in Exhibit 7.1, he criticized it because items 2 and 3 did not allow for problems that were not formulated by the teacher. It is clearly desirable that students should be able to formulate their own problems as was the case with the experimental investigations in Engineering Science at the Advanced level. A midway position is to specify the goal but leave the students to determine the experimental procedures with or without limited guidance [24].

Type of guidance	Rule	Solution
1. Expository learning	Given	Given
2. Guided discovery (deductive)	Given	Not Given
3. Guided discovery (inductive)	Not given	Given
4. Discovery	Not given	Not given

Exhibit 7.1: Wittrock's classification of instruction (Wittrock, M. C. (1963)). Verbal stimuli in concept formation: Learning by discovery. *Journal of Educational Psychology*, 54, pp. 183–190.

I asked my students to read Shulman's paper, along with two other items, one of which gave an example of the differences between expository and discovery teaching on designing and making a tool box (Exhibit 7.2). Then I asked them to compare discovery with expository teaching in their normal classes [25]. In later years the students were asked not only to set their tests a fortnight or so after the activity, but to give another test after about 6 weeks had elapsed to see what had been understood.

---

1. Trainee A is told exactly what tools to use, how to use them, is given preformed parts to assemble, told how to assemble them, and is closely checked and corrected by the tutor during the assembly stage.

2. Trainee B is asked to consider various designs for toolboxes and to decide which one is most suitable for the purpose: to select the appropriate tools and materials, to assemble the toolbox according to the chosen design specifications and to refer to the tutor for advice and guidance when problems occur.

Exhibit 7.2: Differences between expository (teacher centered) and discovery (student centered) approaches from Boffy, R. (1985). YTS Core skills and participative learning. *NATFHE Journal*, pp. 20–23.

In the extreme case some of these graduate student teachers interpreted the term expository to mean continuous talking at the class, with little interaction between them and the pupils. The pupils, as might be expected, found those classes boring. About half the student teachers used Gagné's hierarchical model which has been called "guided learning" or "reception learning" (see Journey 10), although most of the reports submitted showed distinctive differences between the two lessons (see Exhibit 7.3).

Wittrock's examples did not always resolve the problem of defining discovery as the illustration in Exhibit 7.4 shows. For example, expository teaching makes use of questioning; surely, if the questioning is in the Socratic form then it is some form of discovery? But, if only some

Aims/objectives	Lesson Phase and Strategy	Content	Questioning
Aims/non behavioral objective. To develop pupils knowledge and understanding of the water cycle.	**Introduction** Expository	(a) Arrange class for working in groups of 5. (b) Tell each group to take out a sheet of paper and appoint a scribe	
Behavioral objective. Pupils will be able to draw and label a diagram of the water cycle.	**Presentation** Guided discovery (group discussions)	(a) Ask each group in turn "where does the water in rivers or the ground come from?" Tell them to write the answer on the bottom of the paper. (b) Check each groups answer ("rain") and ask each group the next question. (c) Repeat the process in (b). (d) Ask groups to connect each element with arrows (they should perceive a cycle at this stage.	Where does rain come from? Where do clouds come from? Where does water vapour (evaporation) come from? Where does water in seas/lakes come from? Where does rivers/ground water come from?
Behavioral objective. Pupils will be able to describe in their own words what happens in the water cycle; in particular they will be able to explain why it is a cycle.	**Application** Guided discovery (reinforcement by written activity)	(a) Label each group an element from the cycle; I am water. Go from group to group (by asking class) to illustrate the cycle. (b) Tell pupils to draw the cycle. (c) Show water cycle on overhead projector. (d) Explain the terms.	
	Conclusion		

Exhibit 7.3: (a) and (b). Lesson plans to compare guided discovery with expository teaching from Donovan I (1992). Part II of The Training of Students in Discovery Methods of Instruction and learning. Monograph 1/92. Department of Teacher Education University of Dublin. (*Continues.*)

Lesson Phases	Lesson Strategies	Content	Questioning
Introduction	Expository	Class Management	
Presentation	Expository	(a) Introduce water cycle by writing title on board. (b) Explain the cycle. Write each element on the board and connect them with arrows. (c) Any questions from the pupils.	
Application	Expository	(a)Display water cycle diagram on overhead projector. (b) Tell class to copy diagram in their notebooks.	
Conclusion	Expository	\Return to topic begun in previous less (pressure, wind etc) (if time)	

Exhibit 7.3: (*Continued.*) (a) and (b). Lesson plans to compare guided discovery with expository teaching from Donovan I (1992). Part II of The Training of Students in Discovery Methods of Instruction and learning. Monograph 1/92. Department of Teacher Education University of Dublin.

students are questioned it can hardly be called a discovery class. Some of the reports of the use of deductive guided discovery left one wondering if the lesson was not expository. Again when work sheets are used it seems they can be used to elicit a discovery mode of thought while others are simply designed to acquire and memorize more information. It was interesting to find that 20% of the students started their expository classes with a "brainstorming" exercise. Perhaps the significant finding was that teachers perceived themselves to be functioning in a different mode even though, in some cases, there seemed to be little difference between the two classes. Many of the graduate student teachers attempted inductive guided discovery but few attempted "pure" discovery strategies. The experimental investigations undertaken by the engineering science students clearly belonged to the "pure" even when limited assistance was given by the teacher [26].

It seems that the pupils found the discovery mode quite demanding, and some resisted this new approach (see Exhibit 7.5). As some of the student teachers reported, "The pupils were

Expository teaching is where one gives the students everything; facts, principles, rules etc. Discovery teaching is where one sets them a problem but guides their efforts along a certain channel which one hopes will help them to discover the solution.

I chose guided discovery as a method because my experience with my students in my teaching practice has been that they rarely arrive where one wants them to be without a considerable amount of help. It has been my practice to refrain as much as possible from telling them what they will find in a particular experiment, in order to provide an element of discovery and the unknown for them. In these situations it is usual for me to have to give considerable guidelines in the interpretation of their results as they are frequently unable to interpret them for themselves.

In the present instance where the content matter involved the lever, my expository group were led up the hierarchy of prerequisites through the concept of the lever, fulcrum, perpendicular distance, moment of force, clockwise and anti-clockwise moments and equilibrium, and the law of the lever. All this was given.

For a pure discovery group I would have given the weights and suspended metre sticks and directed them to see if they could discover any pattern when the stick was balanced with the weights in different positions to be chosen by themselves.

Using the guided discovery approach, I stacked the odds in my favor by directing them to place the weights on one side of the metre stick at specific points. I hoped through this method to have the simple numerical values they would find so obviously related that the law would be apparent after a few examples of equilibrium. Using only two weights, one of which was twice the size of the other, was an attempt to simplify things and stack the odds in my favor.

I confess to finding it difficult to distinguish between the discovery and the guided discovery approach. Although the former is supposed to involve no help from the teacher, by setting up the problem in the first place, he is guiding the learners in a certain direction. If one simply left a meter stick, a piece of string, a retort stand and some weights on the bench and told students to find a law connected with balancing a stick, it is unlikely that they would get very far. One has to guide them to some extent even if it is only in terms of the hidden guidance involved in how you present the problem to them. For this reason, I would prefer to look at discovery methods as a spectrum running from more to less guidance, rather than from an guided/unguided point of view (the description of Kersh's work recognizes this fact by talking of strong and weak discovery conditions).

Exhibit 7.4: From a graduate student teacher's report, illustrating the difficulty he had with defining discovery.

The class was a bit put off when I asked them to find a definition (in science) for themselves and there was a lot of looking around to each other to see what to do, or whether to take instruction. The suggestion that they should come up with a definition was apparently absurd—that's what the textbooks are for and our role is to learn from the books. Thinking isn't popular with this class. I gave everyone a half-meter stick to experiment with, about half the students never really started, they seemed intimidated by their task and, in their charming ways, indicated that I was an idiot to come up with a conclusion.

Exhibit 7.5: Example of Pupils resisting change from a graduate student teacher's report.

required to think." This is not surprising given that much of the teaching they had received was undertaken with a view to helping them remember material for the public examination. One effect of involving the students in their own learning reported in the literature is that they become more motivated [27]. This effect was often reported by the student teachers. Although there was little difference between the scores they obtained from the two tests, the discovery mode enhanced motivation. Laboratory work is easily designed to be student (learner)-centered, but its assessment may be time-consuming (see Exhibit 7.6) [28]. Pre-laboratory exercises may be given. "For example, students presented with a bio-materials problem related to a laboratory activity were asked to generate knowledge through a scientific literature review, synthesize and interpret their findings, and propose a potential solution to the problem" [29].

Experimental investigation. Problem chosen by the students within the syllabus content rules. Is an open-ended investigation in which students are to develop their own lines of inquiry. It is intended to encourage students to devise experimental procedures, to select appropriate apparatus, occasionally to adapt pieces of equipment to new purposes, to perform experiments and to analyze results. It poses an engineering or scientific problem, and involves the student in an analysis of the situation and an appropriate selection of the procedures and techniques for solution. The end point of the particular investigation may or may not be known, but the means for its achievement are comparatively discretionary (N.B. experimental investigations can be simulated by computers).

Exhibit 7.6: Regulations for the experimental investigations required for the Advanced level examination in Engineering Science set by the Joint Matriculation Board, Manchester (See Chapter 3, Heywood, J. (2016). *The Assessment of Learning in Engineering Education. Practice and Policy*. Hoboken, NJ, IEEE/Wiley.

The many papers presented at the 2017 annual conference of ASEE show that engineering educators need have no fears with experimenting with inquiry based learning, and this extends to teams [30]. The problem for many teachers may be that their students are likely to need convincing that the new method will pay-off. But, with the goal of "transfer" clearly demonstrated in the test questions, the learning difficulties that some students have, particularly freshmen may be resolved.

7.5   IS ENGINEERING A DISCIPLINE?

The scholar academic ideology raises several questions for engineering educators that range from the practice of instruction to the curriculum [31]. For example, in order to determine if engineering is a discipline is it necessary to take into account the work that engineers do, or does the discipline exist independently of what they do? Is engineering design a separate discipline, or is it required to make engineering a discipline? An alternative way of stating this question is, "Given that engineering design is a social activity, does it meet the requirements for a discipline as understood by the scholar academic ideology?"

NOTES AND REFERENCES

[1] Schiro, M. (2013). *Curriculum Theory. Conflicting Visions and Enduring Concerns*, 2nd ed., Los Angeles. Sage. 87, 105, 107

[2] Eggleston, J. (1977). *The Sociology of the School Curriculum*. London, Routledge and Kegan Paul. 87, 106

His other two perspectives he calls "reflexive" and "restructuring." The reflexive paradigm is now called "constructivist" and derives from the work of such theorists as Berger and Luckman (i). They argue that knowledge is socially constructed and depends on our experience and environment. In this situation teachers and students should define a curriculum which is real to them in their social context. In this sense the curriculum should be negotiable and worked out to meet the individual needs of students (ii).

In the UK one or two universities have experimented with programmes of independent study where the student negotiates what he/she wishes to do to achieve a degree. This may involve negotiation of the methods of assessment (see Exhibit 7.7). In engineering a student might negotiate the terms of a substantial project rather than being told what to do. This and the investigations submitted for the JMB Engineering Science Examination were negotiated by the student (iii). Most of the final year of a course at the University of Toronto was available for the students to design themselves within the constraints of prerequisites and requirements for accreditation (iv).

Ruthven (v) in a short paper examined the practical implications of the disciplines thesis for curriculum design in mathematics with some surprising conclusions, for it would seem at first sight that mathematics is a subject in which the disciplines thesis is clear cut. On the contrary, Ruthven argues that common sense and social conceptions of the disciplines are in conflict with logical conceptions. "It is," he writes "a contingent social

The Nature of the Beast

Independent studies offer you a different way of making use of the resources available in the University—principally (a) range of people (the academic staff) with mastery of their particular subject area, (b) books, lab facilities etc and (c) other students. The orthodox way is to package these up into courses, and then to allow a student a structured choice amongst them. The independent does not choose amongst packages, but constructs his or her scheme of studies directly out of the raw materials out of which the packages themselves are made. So the challenging but liberating question an intending student faces is; how can I draw on the resources represented by the university in order to get the most out of my two Part II years?

The school invites prospective students to design a scheme of activities with this basic question in mind, and takes the student on if it finds the proposals acceptable.

To stand a good chance of being accepted, proposals have to be "academic" in character, although for good or bad it is often difficult to say what this test rules out. The best guidance is to say that those activities that already figure in established courses in the University are the sort of activities that are likely to be acceptable in an independent studies scheme. The overwhelmingly predominant activity in past schemes has been reading, followed by critical writing, and indeed many schemes have been defined by students simply specifying what reading they planned and under what headings they intend to write in response to what they read. But other activities have also figured: making sculptures, conducting architectural surveys, writing short stories, magazine editing, doing formal logic, play producing, digging (archaeologically), taking photographs, painting pictures, conducting sociological and anthropological fieldwork, etc.

Exhibit 7.7: *Guide for intending students in independent study at the University of Lancaster.* Introductory paragraphs, circa 1984. (Communicated by William Fuge then director of the programme.)

fact, rather than a logical necessity that has led to the tradition of enquiry commonly known as mathematics [....]" Ruthven's perspective is cultural. The reinterpretation of mathematics in strictly logical terms is to ignore the plausibility of the socio-historical context in which it has been taught and developed. The essential question for curriculum design for Ruthven is "How can we reinterpret mathematics so that it will contribute to the development of a rational perspective on the lives and affairs of men?"

Eggleston proposed that the two paradigms could be brought together in a restructuring paradigm—As two related modes of understanding both the realities of knowledge in the school curriculum and the possibility of change therein. The reflexive perspective develops from the received and the restructuring from the reflexive (see Journey 12).

(i) Berger, P. L. and T. Luckman (1966). *The Social Construction of Reality*. London, Allen Lane.

(ii) Boomer, G. (1992). Negotiating the curriculum in G. Boomer et al. (Eds.), *Negotiating the Curriculum. Education for the 21st Century*. London, Falmer Press.

(iii) See Chapter 3 of Heywood, J. (2016). *The Assessment of Learning in Engineering Education. Practice and Policy*. Hoboken, NJ, IEEE/Wiley.

(iv) Smith, H. W. (1994). University of Toronto curriculum in electrical and computer engineering. *IEEE Transactions on Education*, 37(2), pp. 158–168.

(v) Ruthven, K. (1978). The disciplines thesis and the curriculum: a case study. *British Journal of Educational Research*, 26, pp. 163–176.

[3] (i) "Mans essence is his ability to think, to know, to reason, to reflect, to remember, to question, to ponder...." (Schiro, p. 24). 87, 88, 106

(ii) Primarily educational decisions are about the curriculum and teaching. At the level of the curriculum are decisions about the content. Sometimes there are even more fundamental decisions and in the United States there is an argument about whether or not engineering education is a discipline in its own right. This has been slightly confused with the issue as to whether educational research is a discipline and by inference research in engineering education. The fundamental question is what is a discipline? There were arguments among educational philosophers about this in the nineteen sixties and seventies and the purpose of the paragraphs that follow is to examine if they shed light on the nature of disciplines. The debate between two philosophers Hirst (i) an Englishman and Phenix (ii) an American illustrate the point. Both presented philosophical theories that have a bearing on the curriculum. They were published in 1964 and 1965 respectively. Both derive from the Platonic view that the objects in the "sensible" world are manifestations of "ideals" or "prototypes" held in the mind. The sensible world is a world of the "particular" and they belong to the world of *becoming* whereas the *ideas* or *forms* belong to the intellect, which resides in the world of *being*. These forms are organized in a system the top of which is the form of the *good*. Knowledge is of an absolute and permanent order of ideas. For each true universal concept there corresponds an objective reality (iii). True knowledge is therefore of the universal. Knowledge of the universal (e.g., goodness) is the highest kind of knowledge and knowledge of the particular is of the lowest kind of knowledge. Today, for example, we judge the knowledge required of professional engineers to be more universal and abstract than that required by technicians. Thus, the degrees of knowledge are distinguished according to objects, and the human mind develops from opinion to knowledge. Much of what we do in engineering education is based on tradition and opinion about how students learn. Not on knowledge.

The application of this idea can be seen in the work of Piaget (iv) and Kohlberg (v). In Piaget's theory the child moves through stages of concrete operations to formal reasoning which is the highest level of abstraction. For Kohlberg moral development begins with black and white opinions and develops an all-embracing and abstract concept of justice. For Plato progress is neither, continuous or automatic. It requires effort and mental discipline hence the importance of education that is, to bring youths from opinion to the sight of eternal and absolute truths.

Because it is our thoughts that grasp reality so it is that the objects of thought (as opposed to sense perception) have reality and, in order to grasp them we have to *discover* them for they are not of our invention.

In Plato's *Phaedo* (vi) a method is proposed by which the learners can develop toward the ultimate form. It is the method of hypothesis deduction. A starting point (hypothesis) is assumed and the consequences are examined (deduction) with a view to destroying the untruth in hypotheses until the truth is reached. It is easy to see in this theory of knowledge and form the modern case for discovery learning as put forward by Bruner (vii) and others, or the view that education is about enabling others to discover their own potential. It has profound consequences for the content of the curriculum and instruction. Discovery learning in the guided mode has been shown to take longer than information giving teaching for the same concept(s), but it has also been shown to enhance motivation and there is some evidence that it increases understanding of concepts and principles.

Similarly, the idea that there are forms of knowledge that individuals can learn (such as being, identity, difference, motion and rest as they are connected with or cut-off from one another), leads to a view that there may be a particular curriculum for learning to which we should attend if we are to grow in stature or, as Aristotle would say "wisdom."

This account of Plato has been simplified in the extreme in order to illustrate the point that the derivation of the curriculum is a complex activity dependent wholly or in part on the designer's concept of knowledge even if the designer is not aware that this is the case. The positions of Hirst and Phenix in this regard will also be simplified in order to demonstrate the usefulness of philosophical approaches to the design of the curriculum.

It would seem self-evident from observations of human behaviour that in order to think, act and relate with one another there have to be common frames of reference and that somehow our mind must relate to them. It also seems from observation that different modes of thinking are used to solve different kinds of problem. For example the founding Vice-Chancellor of the University of Lancaster—Charles Carter believed that all students in the humanities should study a science, appropriately taught, because it represented a different mode of thinking. The liberally educated person requires too understand different modes of thought. He did not take this to its logical conclusion namely that scientists should study a subject in the humanities. But the same issue applies to the engineering curriculum if it is held that engineering is a component of liberal education. Phenix may be interpreted at this level of particularity to say that the meanings which such persons (scientists, engineers, artists, historians, etc.) give to their experiences differ, and we need to understand that they do. When, however, in the abstract Phenix examines all possible distinctive modes of understanding he concludes that there are six fundamental patterns that engender essential meanings. These are symbolic, empirics, aesthetics, synnoetics, ethics and synoptics. Their relation to the disciplines as we currently understand them is shown in Exhibit 7.3 which has been simplified. They are the foundations for all the meanings that enter into human experience. They are foundations in the sense that they cover the pure and archetypal kinds of meaning which determine the quality of every humanly significant experience. The relationship with the curriculum for general education is both stated and self-evident.

Phenix introduces the idea of competences into his thesis and in this respect there is much similarity with the ideas of *The Taxonomy of Educational Objectives* or the *SCANs* or *REAL* reports. But there is more to it than that for an emotional dimension is evident that is not evident in these other classifications. Phenix says these realms of meaning "may be regarded as comprising the basic competences that general education should develop in every person. A complete person should be skilled in the uses of speech, symbol and gesture, factually well informed, capable of creating and appreciating objects of aesthetic experience, endowed with a rich and disciplined life in relation to self and others, able to make wise decisions between right and wrong and possessed of an integral outlook. These are the aims of education for the development of the whole person" (viii).

The term meaning applies not simply to the modes of logical thinking within each of the realms but to conscience, feeling, inspiration and other such processes. Each realm is defined by four dimensions.

1. Inner experience.

2. Rule, logic and principle.

3. Selective elaboration.

4. Expression.

From the perspective of the curriculum there is no limit to the varieties of meaning but some are less important. So those that are selected have to be capable of growth and elaboration. From the perspective of the engineering curriculum the inclusion of an emotional (taken in the broadest sense) dimension is of considerable importance for it focuses not only on personal development but on the relationships we have with others. It is notable that a major criticism not only of engineering graduates but of graduates in general in the UK is that they lack "people" skills (in the jargon—"personal transferable skills"). Here then is a philosophy of knowledge that leads to a clear statement of the aims of education with which many people will concur. Would that it were so easy.

Hirst takes issue with Phenix because although the meanings derive from a classification of the objects of knowledge Phenix's approach to classifying these objects is confusing. In effect he removes the emotional dimension for Hirst thinks Phenix is *"mistaken in thinking that knowledge must then be taken as a category wide enough to cover existential awareness and other intelligible states"* (ix).

In order to distinguish between the objects of knowledge Phenix classifies propositions by two dimensions. These are quantity (singular, general comprehensive) and quality (fact, form, norm). Apart from criticising the terms within the quality dimension Hirst asks why these two features should have been selected when there are other possibilities. For example propositions may be classified by tense (past, present, future). "Manifestly one can classify propositions in a great variety of ways but if we are to classify them as true propositions and nothing else, we must do this by virtue of their logically necessary features and not by any other characteristics that they may happen to have" (x). This is how we happen to classify concepts. We become confused about concepts if we take into account properties that do not define them. Therefore, argues Hirst the criteria that distinguish the objects of knowledge are: (1) concepts appropriately related in a logical structure so that propositions can be formed and (2) criteria for judging propositions to be true. This point illustrates the importance of concepts in learning and validates all the work that is being done in assessment in engineering to ensure that concepts are understood. It is on this basis that Hirst proposed his own classification of the forms of knowledge (Exhibit 7.4).

Within these areas other important classifications of knowledge have to be recognized. These he refers to as "fields of knowledge." They are held "together simply by their subject matter drawing on all the forms of knowledge that can contribute to them." He cites engineering, geography, legal, political and educational theory as example of fields of knowledge. He uses political, legal and educational theory to illustrate the idea that moral knowledge is a distinct form of knowledge. It does not have to be subdivided since "moral questions because of their character, naturally arise alongside questions of fact and technique, so that there have been formed fields of practical knowledge that include distinct moral elements within them rather than sub-divisions of a particular discipline" (xi). This seems to suggest that moral and ethical education should not be separated from the subject being taught which asks subject teachers to add moral and ethical considerations to their normal classroom teaching. And this in turn begs questions about the adequacy of their training to undertake this role as well as the adequacy of the preparation given.

Hirst is led to classify knowledge into disciplines and fields of knowledge. "It is the distinct disciplines that basically constitute the range of unique ways we have of understanding experience if to those be added the category of moral knowledge" (xii).

The disciplines of knowledge are mathematics, physical sciences, human sciences, history, religion, literature, fine arts and philosophy. The fields of knowledge embrace both the theoretical and practical which may or may not include elements of moral knowledge.

As Hirst points out there are similarities between the forms of knowledge and the realms of Phenix. There is agreement between them about empirics, aesthetics and ethics. However, Hirst does not believe that symbolic, synnoetics and synoptics represent fundamental categories of knowledge. All knowledge and it is difficult to disagree "involves the use of symbols, [and making of judgements in ways that cannot be expressed in words and can only be learnt in a tradition]" (xiii).

Whitfield (xiv) argued that although Phenix presents a less well substantiated set of categories they may be more useful in planning a curriculum, a view Hirst rejected. Hirst insisted that neither the *realms* or, the *forms* provide a pattern of curriculum units. It is hard to disagree with that but it is equally hard not to take the view that they form a fundamental framework from which curriculum decisions can be derived.

Whether or not the schemes of Hirst and Phenix provide the support that traditionalists (subject specialists) require for a received curriculum seems open to doubt, for men and women come to understand the meaning the world has for them through the solution of practical problems (xv). While they may depend on the forms of knowledge to achieve this function it is unlikely they will recognise this to be the case. Moreover since most problems require information from more than one category of knowledge for their solution, they are likely to function within and across fields of knowledge without reference to the forms. We might note the enormous persistence of the disciplines a fact, which surely of itself, lends support to the idea of a limited number of fundamental categories (forms) of knowledge. One critique of these approaches asked, "How are the structures of the disciplines related to how we structure our experience in perception? This involves asking whether the conceptual structures we have are logically prior to, and the only correct means of structuring our thought and experience or whether they are instead convenient and conventional?" The answer to this question has profound implications for curriculum design. However it is unlikely that those responsible for the design of the curriculum will take into account the forms of knowledge. Their ideas about what characterizes an appropriate curriculum are more likely to be determined by the prejudice of history, convenience and the influence these have on their perception of future needs. Nevertheless, the aims of education have a philosophical basis and philosophical analysis of this kind can be very helpful in the determination of those aims even if we disagree with the epistemological principles on which they are based. There is a case that all teachers in engineering should have a philosophy of education (xvi).

(i) Hirst, P. (1975). *Knowledge and the Curriculum.* London, Routledge and Kegan Paul.

(ii) Phenix, P. H. (1964). *Realms of Meaning*. New York, McGraw Hill.

(iii) Copleston writes "If a man is asked what justice is, and he points to imperfect embodiments of justice, particular instances which fall short of the universal ideal e.g., the action of a particular man, a particular constitution or set of rules having no inkling that there exists a principle of absolute justice, a norm and standard, then that man's mind is a state of opinion [….] He sees the images or copies and mistakes them for originals. But if man has an apprehension of justice itself, he can rise above the images to the form, to the idea, to the universal, whereby all particular instances must be judged, then his state of mind is a state of knowledge [….] Moreover, it is possible to progress from one state of mind to the other, to be 'converted' as it were, and when man comes to realise that what he formerly took to be originals are in reality images or copies i.e imperfect embodiments of the ideal…when he comes to apprehend in some way the original itself, then he has been converted to knowledge." Copleston, F. (1946). *A History of Philosophy. Greece to Rome*, vol 1, London, Burn Oates and Washbourne. Page 152.

(iv) See Crain, W. (1992). *Theories of Development. Concepts and Applications*. 3rd ed., Englewood Cliffs, NJ, Prentice-Hall.

(v) Kohlberg, L. and E. Turiel (1971). Moral development and moral education in G. Lesser (Ed.), *Psychology and Educational Practice*. Chicago, Scott Freeman.

(vi) These remarks are based on Copleston see note (iii). Chapters on Plato.

(vii) Bruner, J. (1961). *The Process of Education*. Cambridge, MA, Harvard University Press.

(viii) *loc. cit.*

(ix) *loc. cit.*

(x) *ibid*

(xi) *ibid*

(xii) *ibid*

(xiii) *ibid*

(xiv) Whitfield, R. C. (1971). *Disciplines of the Curriculum*. New York, McGraw Hill.

(xv) This is a controversial view that finds support in the work of the Scottish Philosopher John Macmurray. See Macmurray, J. (1955). *The Self as Agent*. London, Faber and Faber.

(xvi) Heywood, J. (2017). Philosophy and engineering education. Should teachers have a philosophy of education? *Proceedings Annual Conference of the American Society for Engineering Education*. Paper 18362.

[4] Piaget argues that children move through orderly stages of development. The first stage is from birth to about one and a half years. This is the stage of development of sensori-motor intelligence. Within this stage there are six sub-stages. Each of these is a problem solving activity involving its own logic. After 18 months the child is able to solve a detour problem by going round a barrier even if this means departing from the original goal, for a short time. The child can infer causes from the observation of effects, and begins to predict effects from observing causes; the child also begins to invent applications of something previously learned. 87, 109

The second stage of development is called the period of representative intelligence and concrete operations. This takes the child up to about 11 or 12 years. The first part of the period is between 2 and 7 years and is called the pre-operational stage. The second phase is that of concrete operations. It is in this period that the child learns conservation. Piaget claims that the order of such learning is invariable. Learning by doing is the essence of concrete operations. In this period children learn to seriate, experiment, classify, and establish correspondence.

In the final period the child moves to adolescence. It is the stage of formal operations when the child begins to think in abstraction, to hypothesize, deduce, experiment and theorize. It is the stage of in-built maturity.

Piaget, who is in the Platonic tradition, believes that in the mind there is a cognitive know-how which he calls "structure." One of these structures is the logico-mathematical. These structures enable the child to assimilate the external environment. But the *assimilation* of new information also requires that there should be a change in the existing structures so that there is congruence between external reality and the child' mental structures. This process is called *accommodation*. *Equilibration* is the adjustive process required for *assimilation* and *accommodation*.

[5] That the national curriculum is viewed as a preparation for life was confirmed by the Chief Inspector for Schools in a BBC Interview during the 1 pm News (Radio 4) March 10, 2017. 88, 109

[6] Schiro cites the British educational philosopher R. C. Whitfield who wrote: "Initiation into the disciplines of knowledge, our vehicle for becoming fully human is the worthwhile activity for the curriculum of general education. It provides the base upon which the person as a person can develop to realize his full statute as a free mind and as a citizen. All this is not to imply that the individual and society are not important, but they become, temporarily at least, secondary, as we endeavour to establish a framework for selecting kinds of learning experience which will indicate knowledge and abilities of most worth [....] We should therefore ground our curricular objectives in the distinctive disciplines of knowledge, rather than in social needs, theories of personality, or in a national base

knowledge for 'living in a modern world.' For it is the disciplines themselves which pre-determine these important factors, as well as our underlying ethical conception of what is good and what is worthwhile." Whitfield, R. C. (Ed.), (1971). *Disciplines of the Curriculum*. London, McGraw Hill. Page 9. 88, 109

[7] *ibid* 88

[8] *loc. cit.* Ref. [1] page 4. 88

[9] Bruner, J. (1966). *Toward a Theory of Instruction*. Cambridge, MA, Harvard University Press. 88

[10] *loc. cit.* Ref. [1] page 27. 88

[11] *ibid* 88

[12] *ibid* page 46. 88

[13] Schubert, W. H. (1997). *Curriculum: Perspective, Paradigm and Possibility*. Upper Saddle Creek, NJ, Prentice-Hall. Page 83. 89, 105

[14] Bruner, J. (1960). *The Process of Education*. Cambridge, MA, Harvard University Press. 89

[15] Matthews, G. B. (1980). *Philosophy and the Young Child*. Cambridge, MA, Harvard University Press. 89

[16] Stannard, R. (2005). *Black Holes and Uncle Albert*. London, Faber. 89

[17] *loc. cit.* Ref. [13] page 219.

[18] Bruner, J. (2009). Man a Course of Study in Flinders, D. J. and S. J. Thornton (Eds.), *The Curriculum Studies Reader* 3rd ed., New York, Routledge. The article was reprinted from Bruner, J. [19]. 89

[19] *loc. cit.* Note [13]. 90, 105

[20] Shulman, L. S. (1970). Psychology and mathematics education in E. G. Begle (Ed.), *Mathematics Education*. Chicago, IL, Chicago University/Press National Society for the Study of Education. 90, 105

[21] Bruner, J. (1966). *Toward a Theory of Instruction*. Cambridge, MA, Harvard University Press. 90

[22] *loc. cit.* Note [20]. 90

[23] The search engine for the Proceedings gave different information at different times but there were some 20 papers with "inquiry" in the title. There was no mention of discovery in any of them. They are related to present day terminology, as for example, "active learning" (a), "situated learning" (b), "interactive engagement" (c), and more generally "experiential learning" (b). 90

   (a) Edwards, R. and G. Recktenwald (2017). Guided inquiry in an engineering technology classroom. *Proceedings Annual Conference of the American Society for Engineering Education.* Paper 293.

   (b) Kasprzak, E. et al. (2017). Experiential learning in vehicle dynamics education via motion simulation. *Proceedings Annual Conference of the American Society for Engineering Education.* Paper 1120. (New educational material is presented in an authentic context, and social interaction and collaboration are required for learning to occur).

   (c) Ross, R. and J. Ross (2017). Tales from the wave front: Teaching the physics of cell phones and wireless integration. *Proceedings Annual Conference of the American Society for Engineering Education.* Paper 680.

[24] Maarek, J.-M. (2017). Student feedback in inquiry-based laboratories for Medical Electronics course. *Proceedings Annual Conference of the American Society for Engineering Education.* Paper 18667. 90

[25] Heywood, J. (2008). *Instructional and Curriculum Leadership. Towards Inquiry Oriented Schools,* Dublin. Original Writing for the National Association of Principals and Deputies. 91

[26] *loc. cit.* Note [2] (iii) and other chapters. 93

[27] Bannerot, R. (2017). Hands-on projects in an early design course. *Proceedings Annual Conference of the American Society for Engineering Education.* Paper 809. 95

[28] *loc. cit.* Note [3] (iii) and other chapters. 95

[29] Carlsen, R. W. and R. Morris (2017). Effectiveness of incorporating inquiry-based learning into pre-laboratory exercises. *Proceedings Annual Conference of the American Society for Engineering Education.* Paper 19464. 95

[30] (a) Douglas, E. (2017). Guided inquiry lessons for introduction to materials *Proceedings Annual Conference of the American Society for Engineering Education.* Paper 116. 95
   (b) Kussmaul, C. L., Mayfield, C., and Helen, H. Hu (2017). Process oriented guided inquiry learning in computer science. The CS-POGIL and IntroCS-POGIL projects. *Proceedings Annual Conference of the American Society for Engineering Education.* Paper 17928.

Both these papers are based on a learning cycle. See Douglas E. P. and C. Chiu (2013). Implementation of process oriented guided inquiry learning (POGIL) in engineering. *Advances in Engineering Education* 3, 3; and, Boykin, A. W. and P. Noguera (2011). *Creating Opportunity to Learn: Moving from Research to Practice to Close the Achievement Gap.* Alexandria VA, Association for Supervision and Curriculum Development.

[31] Williams, R. (2003). Education for a profession formerly known as engineering. *Chronicle of Higher Education*, issue January 24th. 88, 96

See in particular the discussion in Froyd, J. E. and J. R. Lohmann (2014). Chronological and ontological developments in engineering education as a field of study in A. Johri and B. M. Olds (Eds.), *Cambridge Handbook of Engineering Education Research*. New York. Cambridge University Press. They use Fensham's model as the basis of their argument. Fensham, P. J. (2004). *Defining an Identity. The Evolution of Science Education as a Field of Research*. New York, Springer.

Schiro [1], page 27, writes that three basic descriptions have generally been used to define an academic discipline: (1) a well-defined area of study, (2) the collection of facts, writings and other works of scholars associated with well-defined areas of study, and (3) a "community of individuals whose ultimate task is the gaining of meaning" in one domain of the world of knowledge [32].

[32] King, A. B. and J. A. Brownell (1996). *The Curriculum and the Disciplines of Knowledge*. New York, Wiley. 107

J O U R N E Y   8

Intellectual Development

8.1   THE SPIRAL CURRICULUM

Both Bruner and Piaget are developmental psychologists. In terms of the discussion in the last Journey 7 the essential difference between them relates to the readiness of a person for a particular type of learning. Bruner believed that readiness could be taught in contrast to Piaget who thought it was invariant, That is, the child becomes ready for transition to a particular stage of development at a particular time irrespective of any instruction. The child is ready when the child is ready.

In addition to his theory of instruction Bruner proposed a spiral model of the curriculum that would help students develop skill in abstraction. It has relevance to teaching in engineering. The principles of the spiral curriculum are:

The curriculum is recursive [1]. Key topics (concepts, principles) are revisited on several occasions during the course.

1. Each visit is at a deeper level of abstraction (difficulty).

2. There is feedback between each visit so that what has been learned is reinforced.

3. Competence should increase with each visit.

4. Reflection plays an important role in the activity of recursion.

An engineering curriculum based on a "spiral approach in which concepts were introduced at an applications oriented level, and then repeatedly revisited with greater levels of sophistication" was organized in the classrooms and laboratories of first and second year students at Worcester Polytechnic [2]. Other examples of the use of the spiral curriculum and its role in developing reflective thinking will be found in the literature on medical education [3], as well as engineering [4, 5, 6].

In another example Woods who cites the teaching of quantum mechanics seems to differ from Bruner's view. He allows that not all the concepts required at more complex levels of abstraction have to be included in the simplified approach. But, he cautions that a simplified approach should contain the key concepts of the more complex approach.

He writes: "As an example I wrote a paper in which I demonstrated that the elementary concepts of quantum mechanics could all be taught very simply... technically, this method avoids discussion of Schrödinger's equation instead deriving as much as possible from the de Broglie relation, which is a far more primitive concept… Often, the detailed results of the simplified rules are not the same as those given by the full calculation, and so the full calculation may still need to be covered later by a more advanced course,- (as Schrödinger's equation would need to be if my simplified introduction to quantum mechanics were presented initially). I believe that a teacher should try to ensure that as far as possible, whatever simplified approach is used does contain as many as possible of the essential concepts of the more complex approach, and that it is these (rather than the actual simplified results) that are pointing out, for example, the dependence (or not) of the results on particular parameters introduced in the calculations" [7].

A spiral curriculum is difficult to design, and at school level it is possible that there is no advance in learning with each recursion, to the extent that it is mere repetition. If there is no development in learning, then repetition is time wasting [8]. It is thought that some topics may not be suitable for a spiral curriculum. If a spiral curriculum is to be used then it has to be carefully planned.

The spiral curriculum raises a major issue about the extent to which engineering schools, or better the engineering curriculum relies on what happens within schools. Should certain cognitive (problem solving) skills be developed in schools through a recursive curriculum that continues in higher education?

8.2   ENGINEERING AND THE SCHOOL CURRICULUM

Some years ago I presented a paper at the Frontiers in Education Conference which showed the multiple choice question in Exhibit 8.1 [9]. You might like to try and answer it before reading on.

You may, or may not be surprised to hear that the children engaged in this primary (elementary) school programme ranged in age from 5 to 13. You may, or may not be surprised to hear that the title of the programme was School Children Operating and Organizing a Profitable Enterprise (SCOOPE). The citations are from posters prepared for the schools in this prgramme by the project officer Ann Ryan.

The idea of the sponsors, The Tipperary Leader Group, was that children participating in this scheme would develop and run for profit what is sometimes called a "min-company.." One school, for example, made an audio tape of the school choir singing Christmas music, and sold it in the locality for charity. Another small school in rural Tipperary involved every child in the school in producing a book about the school and the village. I was privileged to be invited to evaluate the SCOOPE project.

The sponsors believed that Ireland was short of entrepreneurs, and that the group had as its function, the development of entrepreneurial attitudes throughout the school system. It took the controversial view that you did not try and achieve this goal through the Transition year, which is a year between the ages of 15 and 16 when, among other things, students do work in firms and in the community as a means of acquiring skills that will prepare them for life and work (Exhibit 8.2). Rather the Leader Group thought the pupil should begin to develop these

Citation 1

"Students design and build an invention of their choice, and explore entrepreneurial topics, including profitability, marketing, raising venture capital, angel investors and patenting. Creating mini-business plans forces teams to estimate the manufacturing cost of their product and forecast potential sales revenue" [2].

Citation II

"The hands-on experience of developing project facilitates the students to develop skills required to operate a business with regard to the generation of ideas, team work research, management, marketing skills, selling, record keeping etc" [3].

Citation III

"Basic steps of brainstorming

*Select a topic e.g. Ideas for business.

*get a group of 6 to 10 persons together.

*Select someone in the group to write down ideas.

*Accept all ideas-even daft ideas can sometimes work.

*Think as many ideas as possible.

*have fun but stay on focus." [4]

Citation IV

The 4 key parts of the marketing plan are product, price, place and promotion.

*Is it the right product, design quality, color etc? Is it what the customer wants?

*What are people willing to pay for your product?

*Make sure your price covers your costs and gives you a profit.

*Where and how is it easiest for your customers to buy from you?

*Tell customers about products or service using word of mouth, posters, advertising, publicity etc [5]

Question

Which pair of the above citations relate to a primary (elementary) school programme?

    A. I and II

    B. I and III

    C. I and IV

    D. II and IV

The correct answer is D. (But III also relates to the primary curriculum).

Exhibit 8.1

Transition, skill and competencies that the student should have experienced by the end of the Transition Year Opportunity (TYO)

- Have been exposed to a broad, varied and integrated curriculum and have developed an informed sense of his/her own talents and preferences in general educational and vocational matters (*transition skills*).
- Have developed significantly the basic skills of literacy, numeracy and oracy (It is assumed that most students will have developed these skills before the end of junior cycle, but specific reinforcement may be needed for some through TYO) (*literacy, numeracy skills*).
- Have developed confidence in the unrehearsed application of these skills in a variety of common social situations (*adaptability*).
- Have experienced as an individual or as part of a group, a range of activities which involve formal and informal contacts with adults outside the broad school context (*social skills*).
- Have developed confidence in the process of decision making, including the ability to seek out sources of support and aid in specific areas (*decision making skills*).
- Have developed a range of transferable thinking skills, study skills and other vocational skills (*learning skills*).
- Have experienced a range of activities for which the student was primarily responsible in terms of planning, implementation, accountability and evaluation, either as an individual or as part of a group (*problem-solving*).
- Have developed appropriate and physical and manipulative skills in work and leisure contexts (*physical*).
- Have helped to foster sensitivity and tolerance to the needs of others and to develop personal relationships *(interpersonal/caring)*.
- Have been enabled to develop an appropriate set of spiritual, social and moral values (*faith; morals*).
- Have had opportunities to develop creativity and appreciation of creativity in others (*aesthetic*).
- Have developed responsibility for maintaining a healthy life style, both physical and mental (*health*).
- Have developed an appreciation of the physical and technological environments and their relationship to human needs in general (*environment*).
- Have been given an understanding of the nature and discipline of science and its application to technology through the processes of design and production (*science; technology*).
- Have been introduced to the implications and applications of information technology to society (*information technology*).

This list of skills and competences is not exhaustive and new ones may emerge through the experience of schools.

Exhibit 8.2: National requirements for school based curriculum development for the transition year between the end of junior cycle and the beginning of senior cycle post-primary (elementary) education in Ireland for students in the age range 15 to 16. (Curriculum and Examinations Board (CEB) (1986). *Transition Year Programmes. Guidelines for Schools*. Dublin, CEB).

skills and attitudes in the primary (elementary) school. They believed, rather like the Jesuits, that permanent attitudes are more likely to be acquired by younger than older children.

It was the intention that the pupils would develop the skill of independent learning, and they would also be allowed the choice of project.

The teachers were much more enthusiastic, and did much more than the guidance notes expected of them. They found it difficult to facilitate and allow the children to choose their own projects. This is sometimes the case in higher education where an accepted goal is also to help students become independent learners. Instructors find it difficult to let go. However, both the teachers and pupils agreed that they had developed skills necessary for teamwork, and especially communication. It seems that the pupils did not all experience all the skills they were supposed to. Teachers and pupils confirmed that this was primarily due to the way the project was organized.

The SCOOPE project was not directly related to engineering. But, it is evident that the students were being invited to go through a process not dissimilar to the design cycle, although in this case they talked the language of business. In a school in Britain the language of design was learned by 9 and 10 year old children.

Also in a school in Britain they created an opera company in four months. Apart from demonstrating the importance of play the investigator. D. Davies asks us to make a considerable jump in our thinking when he suggests that the "thought processes of children and designers may be closer than we have realized" because "successful designers are those who have kept hold of their imaging abilities, and developed them in parallel with other mental attributes. […] My experience with children and a designer together is that they are able to talk the same language, and build on the approaches they hold in common. Because the aspects of the designing activity described above are entirely natural to children they respond instinctively to the apprentice-ship model of education offered by the designer in the classroom rather than to more rigid, curriculum-led attempts to teach children design" [10].

In Canada the discourse of grade 4 children taking a course in "Engineering for Children" was observed by W. M. Roth [11]. He found that prior to the course the students had scant knowledge of engineering and engineering-related techniques. By the end of the course, they had acquired "a competent engineering-related language that allowed them to articulate their experiences. This discourse was striking in its variations and allowed students to integrate their personal meanings" It had come about not through the imposition of textbook definitions but through discourse in their groups. The language they learned was not the result of memory or teacher given definitions. It was "rich engineering design language to talk over and about design artefacts and the activity of designing."

As far as I know there was no follow up study to see if the attitudes acquired through these projects was retained during their second-level education. Sometimes, so it is held, post-primary education can kill-off creativity. For example, some pupils who had transferred to secondary school had told their teacher that they had asked the teacher of business subjects in the secondary school if they could undertake a similar project to SCOOPE. They were told "no." Their primary school teacher suggested that they might undertake a project by themselves, and work from home. This they did. It says a great deal for the impact of the project on some of the pupils. But it raises issues about cognitive development across the curriculum, not only K-12 but through the period of higher education. The central question is, "what happens to that language in post-primary education, and what can be done to preserve it?" In Piagetian terms it is by no means clear that every student arriving at university has successfully negotiated the stage of formal operations.

8.3   CURRICULUM QUESTIONS RAISED BY PIAGET'S THEORY OF COGNITIVE DEVELOPMENT

Piaget's stage theory has had a great influence on teaching. It is argued that students need to be at the stage of formal operations in order to be able to study at university. However, some studies in the U.S. in the 1960's and 1970's suggested that the majority of freshmen in the U.S. had not reached the stage of formal operations when they entered university. These results, however, could have been an artefact of test design [12].

Discussions during these seminars drew attention to the poor performance of freshmen: poor motivation was also highlighted. It was also suggested that as the students moved on-wards and upwards they forgot, or apparently forgot essential material. Three questions arise: (1) "Could it be that some of these students have not reached Piaget's stage of formal operations?" [13] (2) "What were their motivations for studying engineering" or, more pertinently "what were their expectations of the first year curriculum?" (3) Does that curriculum appear to them to be fragmented as Culver found in his studies in the 1970's and 1980's? [14] Or, as some medical schools have found, unrelated to medicine; or in this case their perceptions of engineering? [15].

During the last forty years there has been some recognition that research on adult learning is of value and relevant to higher education [16]. There has been a similar interest in post-Piagetian development in engineering education. In higher education William Perry led the way [17], and in engineering his cudgels were taken up by Culver and Hackos [18].

8.4   INTELLECTUAL DEVELOPMENT: PERRY AND KING AND KITCHENER

Studies of intellectual development provide other insights into the development of critical think-ing that was discussed in Journey 6.

Perry argued that intellectual and ethical development did not stop when a student reached Piaget's stage of formal reasoning. A number of stages follow which if they are not negotiated will limit the student's ability to handle complex (wicked) problems. Exhibit 8.3 shows the model as summarized by Culver and his colleagues. Perry found that freshmen students brought with them attitudes from school that demanded from their tutors black and white right answers which is exactly what the university curriculum should not be designed to do.

Positions 1 and 2: Dualism

All knowledge is known, and it is a collection of information. Right and wrong answers exist for everything. Teachers are responsible for giving information, students are responsible for producing it.

Position 3: Early Multiplicity

Knowledge includes methods for solving problems. There may be more than one right answer. Teacher's help students learn how to learn. Students are responsible for understanding knowledge.

Position 4: Late Multiplicity

Uncertainty with respect to knowledge and diversity of opinion become legitimate. Teachers require evidence to support opinions and design choices, students learn how to think and analyze.

Positon 5: Relativism

All knowledge must be viewed in context. Teachers are consultants. Students can synthesize and evaluate perspectives from different perspectives.

Positions 6 – 9: Commitment with relativism

For life to have meaning, commitments must be made, taking into account that the world is a changing relativistic place.

Exhibit 8.3: Perry's model as interpreted for teaching engineering design by Culver, Woods and Fitch [22].

It is quite easy to check if freshmen are behaving in this way. The problem for the teacher is to promote a class culture that does not have these expectations. Bruner would suggest that Socratic questioning and discovery learning are one way of causing students to begin to change their attitudes.

By stage 3 of Perry's model it is apparent that authority is "seeking the right answers" and only in the future will we know the right answer. In other words, the student learns that there may be more than one answer to a problem. Perry calls these first three stages "dualism." He argues that much teaching, by which he implies the lecture mode, reinforces this kind of thinking.

The student moves on when he/she recognizes that authority does not have the right answers, so from dualism the student moves into a phase of scepticism, for now it is clear that not only does the authority not have the right answers but everyone, including the student, has the right to hold his or her own opinions, and some of these can be supported by evidence. Thus, by stage 5 some answers are found to be better than others and everything has to be considered in context. It is a stage of relativism. Then the student begins to perceive that good choices are possible and that commitments have to be entered into. By stage 9 (acting on commitment) decisions are made with relative ease, a sense of identity and personal style is obtained, and one is able to take responsibility for one's own actions.

If students are enabled to break out from dualism and attain stage 5 much will have been achieved. Clearly, the design of assessment is crucial if deep as opposed to surface learning is to be achieved [19]. Students are so often driven by how they perceive the requirements of faculty for assessment.

There have been a small number of studies of the use of the Perry model in engineering. My review of them led me to the view that first year students are attached to the idea of "real" engineering, and therefore, the importance of design in the freshmen year cannot be over emphasized. While a lot of freshmen courses have been redesigned and probably meet this need, it seems that if students are not motivated, that which was gained in the first year will be lost in subsequent years. It needs to be remembered that intellectual growth is not linear [20]. Student development has to be seen as a "whole" department activity which will undoubtedly require a substantial change in attitude on the part of many of its members.

A model that has many similarities with the Perry model is the Reflective Judgment model of King and Kitchener. I find it very attractive because of the ideas it gives teachers about the teaching of the seven stages (see Exhibits 8.4 and 8.5).

Stage	Description
Stage 1	Knowing is limited to single concrete observations. What a person observes is true.
Stage 2	Two categories for knowing: right answers and wrong answers. Good authorities have knowledge; bad authorities lack knowledge.
Stage 3	In some areas, knowledge is certain and authorities have that knowledge. In other areas knowledge is temporarily uncertain. Only personal beliefs can be known.
Stage 4	Concept that knowledge is unknown in several specific cases leads to the abstract generalization that knowledge is uncertain.
Stage 5	Knowledge is uncertain and must be understood within a context thus, justification is context specific.
Stage 6	Knowledge is uncertain but constructed by comparing evidence and opinion of different sides of an issue or across contexts.
Stage 7	Knowledge is the outcome of a process of reasonable inquiry. This view is equivalent to a general principle that is consistent across domains.

Exhibit 8.4: Stages of the King and Kitchener reflective judgment model.

The model focusses on the development of reflective judgement; reflective thinking is at the heart of critical thinking [21]. The authors developed a reflective judgement interview

Promoting Reflective Thinking Stage 3 Reasoning

Characteristic Assumptions of Stage 3 Reasoning

Knowledge is absolutely certain in some areas and temporarily uncertain in other areas.

Beliefs are justified according to the word of authority in area of certainty and according to what "feels right" in areas of uncertainty.

Evidence can be neither evaluated nor used to reason to conclusions.

Opinions and beliefs cannot be distinguished from factual evidence.

Instructional Goals for Students

Learn to use evidence in reasoning to a point of view.

Learn to view their own experiences as one potential source of information but not as the only valid source.

Promoting Reflective Thinking: Stage 6 Reasoning

Characteristic Assumptions of Stage 6 Reasoning

Knowledge is uncertain and must be understood in relation to context and evidence.

Some points of view may be tentatively judged as better than others.

Evidence on different points of view can be compared and evaluated as a basis for justification.

Instructional Goals for Students

Learn to construct one's own point of view and to see that point of view as open to re-evaluation and revision in the light of new evidence.

Learn that though knowledge must be constructed strong conclusions are epistemologically justifiable.

Exhibit 8.5: Promoting reflective thinking in the King and Kitchener model. Steps 3 and 6. Adapted from King, P. M and K. S. Kitchener (1994). *Developing Reflective Judgement.* San Fransisco, Jossey-Bass pp.251 and 254, respectively. Each tabulation also included sections for difficult tasks from the perspective of the stage, sample developmental assignments, and developmental support for instructional goals.

(RJI) for the purpose of measuring skills associated with reflective thinking. They claim that it measures a construct that is different from the traditional constructs of critical thinking, verbal reasoning, and formal operations. These, they argue, assess problem solving of well-structured rather than ill-structured problems which the Reflective Judgment Interview (RJI) is designed to assess. At the same time they suggest that learning to solve well-structured problems may be a pre-requisite for learning the higher level skills they perceive to be demanded by reflective judgment. King and Kitchener say that teachers need to "ask whether their students have the inductive and deductive skills that appear to be necessary prerequisites for higher levels of reflective thinking" (p. 202).

As indicated the model has many similarities with the first three or four stages of the Perry model, and may be describing the same things. The stages or levels of the model are representative of pre-reflective thinking, quasi-reflective thinking (stages 4 and 5), and reflective thinking (stages 6 and 7). (See Exhibit 8.4.)

The RJI typically consists of four ill-structured questions that focus on the concepts of the model. "The four standard problems concern a range of issues: how the Egyptian pyramids were built, the objectivity of news reporting, how human beings were created, and the safety of chemical additives in food." They have also used the problem of nuclear waste.

They describe a problem from two contradictory points of view with the purpose of studying how persons' reason about the intellectual issues involved. The interview is semi-structured. After the question has been read out a series of probe questions are asked; they might be followed up by other questions in order to focus or clarify, or refocus the response (p. 102). The RJI has been criticized for being gender biased. It should not be assumed that the stages of the model are fixed and some-how related to the structure of the curriculum [22].

Moore and Hjalmarson [23] reported that given an appropriate learning environment using an MEA first-year engineering students showed that they were capable of working on complex problems [24] (see Section 6.4).

Finally, Exhibit 8.6 records engineer and educator John Cowan's levels of reflective thinking, and is a reminder that his book on reflection in action in higher education is an excellent source of reference for the beginning engineering educator [25].

NOTES AND REFERENCES

[1] W. E. Doll Jr. presented a four criteria alternative to Tyler's model of the curriculum. These criteria were richness, recursion, relations and rigor. He argued that recursive reflection lied at the heart of any transformative curriculum. Dewey, Piaget and Whitehead all advocated the process. Although Bruner's MACOS (Man a Course of Study) failed, the spiral curriculum was an attempt to define and describe a recursive curriculum. "Recursion and repetition differ in that neither one, in any way, reflects the other. Repetition, a strong element in the modernist mode, is designed to improve set performance. Its frame is closed. Recursion aims at developing competence-the ability to organize, combine inquire, use something heuristically. Its frame is open. The functional difference between repetition and recursion lies in the role reflection plays in each. In repetition reflection plays a negative role; it breaks the process. There is a certain automaticity to repetition that keeps the same process going-over and over and over, as in a flash card, arithmetic drills. In recursion, reflection plays a positive role; for thoughts to leap back on themselves [...] it is necessary as Bruner has said, to step back from one's doings to 'distance oneself in some way' from one's own thoughts. Thus, in recursion it is a necessity to have

Levels of Reflective Thinking	Examples of thinking that characterize the levels
A. Summarizing	What is worrying me most, and why?
B. Analyzing	What conclusions can I draw from what I have thought already?
C. Closing the review	What points emerge from these reflections that are likely to be of use to me?
D. Distinguishing	What questions here might open up reflections that could be valuable to me?
E. Reasoned selection	Which is the best option for me at this moment?
F. On-going evaluation	Have I considered all that I should have thought about?
G. Self-awareness	Have I let myself be unduly influenced by personal preferences when choosing what to do?
H. Creative thinking	What will be the best way to do this?
I. Free thought	Wait a minute, wait a minute. I think I feel another blue flash coming in…..

Exhibit 8.6: John Cowan's nine levels of reflection, each accompanied by one of the several questions he uses to illustrate each level. (Personal communication.)

others […] look at, critique, respond to what one has done." Doll, W. E. (2009). The four R's—An alternative to the Tyler rationale in D. J. Flinders and S. J. Thornton (Eds.), *The Curriculum Studies Reader*. New York, Routledge Falmer. Page 270. 109

[2] Cyganski, D., Nicolleti, D., and J. A. Orr (1994). A new introductory electrical engineering curriculum for the first year student. *IEEE Transactions on Education*, 37(2), pp. 171–177. 109, 122

[3] Hargreaves, K. (2016). Reflection in Medical Education. *Journal of University Teaching and Learning Practice*, 13(2). http://ro.uow.edu/jutlp/vol13/iss2/6 109

[4] Yost, S. and M. Krishnan (2017). Development of an integrated spiral curriculum in electrical and computer engineering. *Proceedings Annual Conference of the American Society for Engineering Education*. Paper 210.

[5] Mallikarjunan, K. et al. (2017). A Spiral curriculum approach to the implementation of instrumentation in biological systems engineering. *Proceedings Annual Conference of the American Society for Engineering Education*. Paper 2065.

[6] Mallikarjunan, K., Whysong, C., and J. Lo (2017). Improving ethics studies through a spiral curriculum: Piloting an ethics discussion at the senior level. *Proceedings Annual Conference of the American Society for Engineering Education.* Paper 2129.

[7] Woods, R. C. (2000). Simplifying the syllabus. Can we avoid throwing out of the baby? *Engineering Science and Education Journal*, February 2, p. 3 cited in Heywood, J. (2005). *Engineering Education. Research and Development in Curriculum and Instruction.* Hoboken, NJ, IEEE/Wiley. 110

[8] Miwa, T. (1992). School mathematics in Japan and the U.S. Focussing on recent trends in elementary and lower secondary school in I. Wirszup and R. Strait (Eds.), *Developments in School Mathematics Education around the World.* Reston, VA, National Council of Teachers of Mathematics. 110

[9] Heywood, J. (2002). SCOOPE and other primary (elementary) school projects with a challenge for engineering education. *ASEE/IEEE Proceedings Frontiers in Education Conference*, F2C-6 to 10. 110

[10] Davies, D. (1996). Professional design and primary children. *International Journal of Technology and Design Education*, 6, pp. 45–59. 113

[11] Roth, W. M. (1996). Learning to talk engineering design: Results from an interpretive study in a grade 4/5 classroom. *International Journal of Technology and Design Education*, 6, pp. 107–135. 113

[12] Reviewed in Heywood, J. (2000). *Assessment in Higher Education. Student Learning, Teaching, Programmes and Institutions.* London, Jessica Kingsley. Page 170 ff. 114

[13] Duncan-Hewitt, W. C. et al. (2001). Using developmental principles to plan design experiences for beginning engineering students. *ASEE/IEEE Proceedings Frontiers in Education Conference*, 1, T3E-17 to 22. 114

[14] (a) Culver, R. S. and J. T. Hackos (1982). Perry's model of intellectual development. *Engineering Education*, 72(2), pp. 221–226. 114, 121
(b) Culver, R. S. and P. Fitch (1990). Design of an engineering course based on developmental instruction. *ASEE/IEEE Proceedings Frontiers in Engineering Education Conference*, pp. 628–630.
(c) Beston, W., Fellows, S., and R. S. Culver (2000). Self-directed learning in an ASI course. *ASEE/IEEE Proceedings Frontiers in Education Conference*, 1, T3G-1 to 6.

[15] Notwithstanding the changes in education made as a result of the coalitions in the U.S. in the 1990's. 114
It is worth reading Flammer, G. H. (1972). Applied motivation. A missing role in teaching. *Engineering Education*, 62(6), pp. 519–522. He wrote "so long as we are obsessed

with content we will never meet the more basic foundation requirements necessary for high level performance after graduation. We just don't have the time to get the student into higher forms of learning activities which would be such effective motivators. And yet I am intrigued with the idea that once we overcome a student's cumulative ignorance by mastery level performance and that once he is 'turned on' he can absorb more content and with real understanding." Hence his advocacy of a self-paced proctorial system of instruction to foster motivation.

[16] (a) Alexander, C. N. and E. J. Langer (Eds.), (1990). *Higher Stages of Human Development.* New York, Oxford University Press. 114
(b) Hoare, C. (Ed.), (2006). *Handbook of Adult Development and Learning.* New York. Oxford University Press.

[17] Perry, W. G. (1970). *Forms of Intellectual and Ethical Development in the College Years.* New York, Holt, Reinhart and Winston. 114

[18] *loc. cit.* note [14](a). 114

[19] Marton and Säljö (i) distinguished between deep and surface approaches to understanding. They concluded that the strategies found, apart from anything else, were indicative of different perceptions of what is wanted from learning. From the perspective of assessment design there is no guarantee that a student will perceive the demands of assessment in the same way as the tutor. Moreover the design of assessment can cause either deep or surface learning. 116

Entwistle and Ramsden (ii) developed an Approaches to Study Inventory that yielded four factors. *Meaning orientation* had high loadings on the deep approach associated with comprehension learning and intrinsic motivation: whereas the *reproducing orientation* was highly loaded on the surface approach, operation learning and improvidence were associated with fear of failure and extrinsic motivation. *Nonacademic orientation* related to disorganized approaches to study, and *achieving orientation* to strategic approaches to study. Previously Entwistle, Hanley and Ratcliffe (iii) had identified a strategic approach to learning in which students try to "manipulate the assessment procedures to their own advantage by careful marrying of their efforts to the reward system as they perceive it."

(i) Marton, F. and R. Säljö (1984). Approaches to learning in F. Marton, D. Hounsell, and N. J. Entwistle (Eds.), *The Experience of Learning.* Edinburgh, Scottish Academic Press.

(ii) Entwistle, N. J. and P. Ramsden (1983). *Understanding Student Learning.* London, Croom Helm.

(iii) Entwistle, N. J., Hanley, M., and G. Ratcliffe (1979). Approaches to learning and levels of understanding. *British Educational Research Journal,* 5, pp. 99–114.

[20] (a) Heywood, J. (2005). *Engineering Education. Research and Development in Curriculum and Instruction*. Hoboken, NJ, IEEE/Wiley. 116

(b) See note [2](b) and also Marra, R. M., Palmer, B., and T. A. Litzinger (2000). The effects of a first year design course on student intellectual development as measured by the Perry scheme. *Journal of Engineering Education* 89(1), pp. 39–45.

(c) For a college approach to the design of a curriculum for development, see Mentkowski, M. and associates (2000). *Learning that Lasts. Integrating Learning, Development in College and Beyond*. San Fransisco, Jossey-Bass.

[21] King, P. M. and K. S. Kitchener (1994). *Developing Reflective Judgment*. San Fransisco, Jossey Bass. 116

[22] Pascarella and Terenzini in their review of research on the RJI found that differences between freshmen and senior scores on the RJI could not be explained by differences in academic ability although it was possible that net effects were confounded by age. This does not null the RJI as a source of ideas for questioning at the higher levels of reasoning. 115, 118

Pascarella, E. T. and P. T. Terenzini (2005). *How College Affects Students. A Third Decade of Research*, vol 2, San Fransisco, Jossey-Bass.

[23] Moore, T. J. and M. A. Hjalmarson (2010). Developing measures of roughness: Problem solving as a method to document student thinking in engineering. *International Journal of Engineering Education*, 26(4), pp. 820–850. 118

[24] Culver, R. S., Woods, D., and P. Fitch (1990). Gaining professional expertise through design activities. *Engineering Education*, 80(5), pp. 533–536. 118

[25] Cowan, J. (2006). *On Becoming an Innovative University Teacher: Reflection in Action*. Maidenhead, SRHE/Open University Press. 118

JOURNEY 9

Organization for Learning

9.1   INTRODUCTION

Substantial criticisms both philosophical and psychological were made of discovery learning as well as the general position that Bruner took. The intention of this chapter is to examine the position taken by David Ausubel in that debate.

Shulman in his review of the work of Ausubel, Bruner and Gagné states Ausubel's position thus, "Ausubel strongly rejects the notion that any kind of process, be it strategy or skill, should hold priority among the objectives of education. He remains a militant advocate of the importance of mastering well-organized bodies of subject-matter knowledge as the most important goal of education." [1].

It is a different view of knowledge to Bruner, and derives from a different view of the objectives of education. It places Ausubel firmly in the group of scholar academics who regard exposition by the teacher as the vehicle for learning. But, what matters for Ausubel is the way in which that exposition is organized. Because Ausubel's and Bruner's objectives of education are different it is difficult to assess "the relative potencies of the theories they espouse," so argues Shulman [2].

Ausubel's approach is discussed here in order that beginning engineering educators can evaluate if his concept of the "advanced organizer," and more generally the concept of cognitive organization has meaning for them.

9.2   THE "ADVANCED ORGANIZER"

Advanced organizers are a form of mediating response. Their intention is to facilitate meaningful learning. As the term "advanced" suggests, they precede a major learning task, although they are a learning task themselves. Ausubel uses concepts and principles to help further explanation, as well as the organization of a more substantial body of material on the same subject. In this way both readiness and structure are provided. Thus, one should precede a lecture with a micro lecture that covers the key concepts and principles to be explained in the main part of the instruction. In short, the principle function of the organizer is to bridge the gap between what the learner already knows and what the learner needs to know before he/she can successfully learn the task at hand. The "advanced organizer" is a sorting and classifying mechanism [3]. It highlights the importance of prior knowledge in bringing about any kind of change (see below).

Advanced organizers contribute to learning in two important respects. First, given that learning is that process by which experience develops new and reorganizes old concepts: the organizers provide a link between the old and the new. Second, advanced organizers contribute to the development of skill in transfer to new learning by providing meaning, structure, and organization about that which is to be learned.

9.3   USING "ADVANCED ORGANIZERS"

Advanced organizers are a sophisticated form of getting a lesson off to a good start. It takes graduate student teachers somewhat more time to learn that it is a good idea to tell the students where they are going, that is, to tell the students the objectives of the course. While there are instances where there might be good reason for not stating objectives, as Eisner has suggested, it seems that learning is enhanced if students know where they are going. This is particularly so when the teacher relates it to specific instructions on how to organize specific information [4]. It is the first stage in constructing a scaffold on which the students can build their understanding of concepts and principles.

Some teachers come close to constructing an advanced organizer when they begin their lesson with a description or survey of what they intend to cover. Many of my graduate student teachers reported that they had used the idea of advanced organization, but the examples I collected suggested that they found them difficult to design, as did I! What they do is commonly practised, and they find the term "advanced organizer" useful in explaining the "what" and "why" of their actions. In this respect, Scandura and Wells definition that an advanced organizer is "in general a non-technical overview or outline in which the non-essential materials of the to-be-learned material are ignored" [5] seems to describe these graduate student-teachers understanding of this technique. It may be argued that a clear statement of outcomes fills this need.

In respect of mathematics, Orton [6] said that the concept of the advanced organizer is valuable even though there are few occasions when new knowledge cannot readily be linked to existing knowledge. Such is the nature of linearity in the subject of mathematics. It will almost certainly be argued by some teachers that this applies to the teaching of engineering science. So, what is the benefit of the organizer, if at all? First, the preparation of an advanced organizer should aid the preparation of the lesson plan. Second, it should show what is important and not important. In these circumstances it should help sort out the necessary from the unnecessary. Jim Stice reported that designing his engineering classes by objectives greatly helped him to eliminate unnecessary material (see Journey 4). A reasonable working assumption is that a learner can only hope to understand 5 or 6 major items in a single fifty minute lecture, if that.

9.4   PRIOR KNOWLEDGE; MEMORY

The importance of prior knowledge cannot be over emphasized. It applies in all walks of life. You will not get change unless the institution or persons you wish to change has some knowledge that can help them assimilate the new knowledge that is required. In that way they should be better prepared to engage in reasoned argument [7].The need for prior knowledge and how to provide it, more especially through problem based, project based, and case studies, may be inferred from recent publications [8, 9, 10].

One of my graduate student teachers reported that, "I had begun human (reproduction) biology and was about to teach the digestive system. I realized that all the books would be using terms like enzymes and starch, proteins and vitamins. Before introducing digestion we had a lesson on diet and the seven types of food used in daily diet. This was a form of advanced organizer and I used it to facilitate the learning of the digestive system. It was a stepping stone to digestion."

In this sense a course may be considered to be a series of advanced organizers. Prawat has pointed out that the use students make of prior knowledge is dependent on the way knowledge is organized in their memory systems [11]. Lessons have to be designed in sequence if students are to construct an adequate scaffold. In these circumstances the advanced organizer facilitates the memory because it draws attention to the specific ideas and principles to be considered [12]. Hence, the importance of key concepts (see Journey 10) for both instruction and curriculum design.

9.5   COGNITIVE ORGANIZATION

It is evident that there are many ways of organizing knowledge. These and advanced organizers are sometimes grouped together under the generic heading of cognitive organization (or organizers). For example, in my 1982 book on teaching the chapters were organized around examination questions, each of which required from 45 minutes to 1 hour to answer. The questions were designed to draw out significant principles that were then discussed and illustrated in the text (e.g., Exhibit 9.1). Studying questions of this kind is valuable since it helps students prepare for examinations by helping them to construct scaffolds so as to prevent them from memorising answers to questions they may predict will appear in the examination paper. The questions should be designed to foster transfer to different situations not directly considered in the instruction. In this way students learn to integrate the disparate knowledge that is put before them in their courses. Students can also be asked to design questions that elicit principles, and for that matter design their mark schemes [13].

Unfortunately, too often we do not consider the design of questions to be an art. Consider the example in Exhibit 9.1. The first question insists that the respondents give examples from their own teaching, whereas the second question would induce examples from a text book. Some of my colleagues would say that an answer to the first question could be fudged, in which case

Question 1. Describe two kinds of advance organizer for use in learning the same material in the subject you teach. Indicate the characteristics of those advanced organizers and say why they will facilitate the acquisition of new information.

Question 2. Construct two kinds of advanced organizers for use in learning the same material. Indicate the characteristics of the advanced organizers and why you think they will facilitate the acquisition of new material.

Exhibit 9.1

you would have to ask the students to submit their teaching journals which would show whether or not they had used an advanced organizer. If the answer is in their journal (portfolio) then you are left with the question, "What is the point of the exam?" And that, in its turn, raises the question, "how is the understanding of the application of theory into practice to be measured?"

In project work it is found that students who can clearly define what it is they want to do at the beginning of the exercise have a much better chance of completing the project than those students whose focus is ill-defined at the beginning [14]. In other words, those who perform well are able to create their own advance organizers, which is what a project plan is. Those in difficulty require much help. This is consistent with the finding that mature students derive greater benefits from exposure to new information and experience than immature students who are unable to apply rules and are inhibited by a limited knowledge base [15].

Cognitive organizers of whatever kind help students to alter and to focus on what is to be learned. They mediate between what has been learned and what is to be learned. They are knowledge dependent in two ways—first, if the students do not have prior knowledge they will be "out of their depth." It is the experience of beginning teachers that getting instruction right is sometimes difficult. If it is not aligned with prior knowledge it may cause minor disruptive behavior. Sometimes a class may ramble along quite cheerfully for several weeks before the beginning teacher finds out that the principles have not been understood. I suspect that many teachers are unaware that this may be happening; I do not differentiate between high school or university teachers!

Second, cognitive tasks that require the active transformation of knowledge are very demanding [16]. It is evident that such demanding tasks can impact on low achieving students. Cognitive tasks should, therefore, be designed with care so as not to reduce the confidence with which students approach their learning.

9.6   MEDIATING RESPONSES

Teachers understand that they have to provide connections between that which has been learned and that which is to be learned through the association of one idea with another. They do this through mediating responses. Parents do this when they help their children to learn through the use of terms such as:

-is like

-is different from.

By using statements like these, particularly those that relate to similarities and differences, new phenomena can be related to the experience of the child. But mediating responses have to be used with care; they may cause ambiguity, or they may not be understood. For example, teachers often use the word *familiarity* instead of *experience*, but they also use *familiarity* in two senses [17]. Great care must be exercised in the choice of such terms to ensure that they are clearly understood by the students in the context in which they are used. Mediating responses like "is different from" may also cause the teacher considerable problems since students may not grasp the significant differences. A teacher is likely to be helped by the research which has been done on concept learning which will be discussed in Journey 10.

One of the important lessons of analytic philosophy is that irrespective of level of intellect and maturity we have to be clear about the meaning of the language we use if there is to be correspondence between our meaning and that which is perceived by students. This applies as much in university lecture theatres as it does to elementary school classrooms. Similarly, much care needs to be taken in the preface to instruction if that instruction is to be meaningful to the students.

9.7   IMPACT OF K-12 AND CAREER PATHWAYS

As will be shown in Journey 10 students often bring misconceptions with them that have been learned before they come to university. This is not surprising we are surrounded with engineered artefacts ranging from bridges to I phones about which judgments are made about how they were designed and work. Therefore, those responsible for teaching STEM subjects in schools have an obligation to ensure that their students understand the concepts [18], and find strategies to influence student learning, as for example through project work [19].

Complaints about the lack of interest in engineering among high school students are the concern of engineering educators in several countries and numerous studies have been undertaken over the years to try and understand the problem. One intervention in the U.S. has been the design of first year programmes that help students to make informed choices. They are programmes that provide prior-knowledge. One study has shown that the reasons students choose to pursue a particular engineering discipline are very field specific. The American system is in contrast to the UK where students directly enter into a specialist field. In this case, there is an onus on high schools to ensure that there is appropriate career guidance. Part of the purpose of career guidance is to provide students with adequate prior knowledge.

NOTES AND REFERENCES

[1] Shulman, L. S. (1970). Psychology and Mathematics education in E. Begle (Ed.), *Mathematics Education*. 69th Year book of the National Society for the Study of Education. Chicago, Chicago University Press. 123

[2] *ibid* 123

[3] Ausubel, D. P. (1968). *Educational Psychology. A Cognitive View.* New York, Holt, Reinehart and Winston. 123

Ausubel writes:

"Logically meaningful material becomes incorporated most readily and stably in cognitive structure in so far as it is subsumable under specifically relevant existing ideas. It follows, therefore, that increasing the availability in cognitive structure of specifically relevant subsumers—by emplanting suitable organizers—should enhance the meaningful learning of such material" [p. 137].

He lists the components of the rationale as follows:

(a) the importance of having relevant and otherwise appropriate established ideas already available in cognitive structure to make logically meaningful new ideas potentially meaningful and to give them stable anchorage; (b) the advantages of using the more general and inclusive ideas or subsumers (namely, the aptness and specificity of their relevance, their greater inherent stability, their greatest explanatory power, and their integrative capacity); and (c) the fact that they themselves attempt both to identify already existing relevant content in cognitive structure (and to be explicitly related to it) and to indicate explicitly both the relevance of the latter content and their own relevance for the new learning material. In short, the principle function of the organizer is to bridge the gap between what the learner already knows, and what he needs to know before he can successfully learn the task at hand [p. 138].

[4] Corno, L. (1981). Cognitive organizing in classrooms. *Curriculum Inquiry*, 11, pp. 359–377. 124

[5] Scandura, J. M. and J. M. Wells (1967). Advanced organisers in learning abstract mathematics. *American Educational Research Journal*, 4, pp. 295–301. 124

[6] Orton, A. (1992). *Learning Mathematics. Issue, Theory and Practice*, 2nd ed., London, Cassell. 124

[7] Heywood, J. (2006). Factors in the adoption of change: Identity, plausibility and power in promoting educational change. *ASEE/IEEE Proceedings Frontiers in Education Conference*, T1B-9 to 14. 125

[8] Spezia, C. (2017). Instructional development and assessment of a task-oriented senior level data acquisition project in a simulated business environment. *Proceedings Annual Conference of the American Society for Engineering Education*. Paper 940. 125

[9] Lee, W. and F. Mak (2017). A Case study: A new course on engineering project and management for first-year graduate students in electrical and computer engineering. *Proceedings Annual Conference of the American Society for Engineering Education*. Paper 635. 125

[10] Delatte, N. et al. (2017). Assessing the impact of failure case studies on the civil engineering and engineering mechanics curriculum: Phase II. *Proceedings Annual Conference of the American Society for Engineering Education*. Paper 531. 125

[11] Prawat, R. S. (1989). Teaching for understanding: Three key attributes. *Teaching and Teacher Education*, 5, pp. 315–328. See also p. 100 of Brown, P. C., Roediger, H. L., and M. A. McDaniel (2014). *Make it Stick. The Science of Successful Learning*. Cambridge, MA, Belknap Press. 125

[12] Gage, N. L. and D. C. Berliner (1988). *Educational Psychology*. 4th ed., Boston, Houghton Mifflin. 125

[13] Heywood, J. (1978). *Examining in Second Level Education*. Dublin, Association of Secondary Teachers Ireland. p. 80. 125

[14] Carter, G., Heywood, J., and D. T. Kelly (1986). *Case Study in Curriculum Assessment. GCE Engineering Science (Advanced)*. Manchester, Roundthorn Press. 126

[15] Brown, A. L., Campione, J. C., and J. C. Day (1981). Learning to learn. On training students to learn from texts. *Educational Researcher*, 10(2), pp. 14–21. 126

[16] Corno, L. (1986). Self-regulated learning and classroom teaching. Paper at annual meeting of the American Educational Research Association cited in Bellon, J. J., Bellon, E. C., and M. A. Blank (1992). *Teaching from a Research Knowledge Base. A Development and Renewal Process*. New York, (Merrill) Macmillan. 126

[17] McDonald, F. J. (1968). *Educational Psychology*. Belmont, CA, Wadsworth. 127

[18] Brophy, S. P. and G. Mann (2017). Teachers noticing engineering in everyday objects and procedure. *Proceedings Annual Conference of the American Society for Engineering Education*. Paper 2535. 127

[19] Oswald, N., Huddleton, C., and A. Cheville (2017). A Race-car design-built-test project for low income first generation pre-college students. *Proceedings Annual Conference of the American Society for Engineering Education*. 127

JOURNEY 10

Concept Learning

10.1 ROBERT GAGNÉ

Robert Gagné also takes an opposite view to Bruner but unlike Ausubel his emphasis, is like Bruner on process rather than product. His theory is in the "objectives" tradition and begins with the question, "What is it you want the learner to be able to do?" Gagné calls this a "capability." In order to achieve this capability it needs to be analysed, and this will produce a hierarchy of tasks that have to be completed in order [1].

In his earliest model the understanding of principles was preceded by the understanding of concepts. A principle was the linking of two concepts e.g., "Birds Fly." Unfortunately, it is somewhat more complex than this for some concepts embrace principles. They become "fuzzy" when there is a debate about the principles that contribute to their structure. The concept of "democracy" serves to illustrate this point. This does not null the theory but is a reminder of the complexities that arise at the higher levels of learning including engineering.

The revised version of his categories of human learned capabilities is shown in Exhibit 10.1. In his first scheme he did not distinguish between the two types of concepts. Of the revised categories he wrote: "Learners have acquired 'concrete' concepts when they can identify previously unencountered instances of a class of objects, a class of object properties, or class of events by instant recognition"[…].

---

1. Verbal information
2. (i) Discrimination
   (ii) Concrete concepts
   (iii) Defined concepts
   (iv) Rules
   (v) Higher order rules (problem solving)
3. Cognitive strategies
4. Attitudes
5. Motor skills

Exhibit 10.1: Gagné's five categories of human learned capabilities.

"Learners have acquired a 'defined' concept when they use a definition to put something they have not previously encountered or put somethings into classes"… Using the term "rule" rather than principle, learners have understood the rule "when they can demonstrate its application to previously unencountered instances." This is what is meant by "transfer of learning."

"Principles or rules derive from relationships between concepts." "Higher order rules" as Gagné now calls "problem solving" are obtained "when two or three more previously learned rules are used to answer a question about an unfamiliar situation [2]."

Cognitive strategies are internal mechanisms for improving the effectiveness of learning. Attitudes are predispositions which shape a person's behavior toward artefacts, events, and people: motor skills are involved in the performance of a physical task.

The attempt to model Gagné's first published scheme shown in Exhibit 10.2 was made by a graduate student teacher of economics. Another graduate student devised the scheme shown in Exhibit 10.3 for algebra for 12 to 13 year olds.

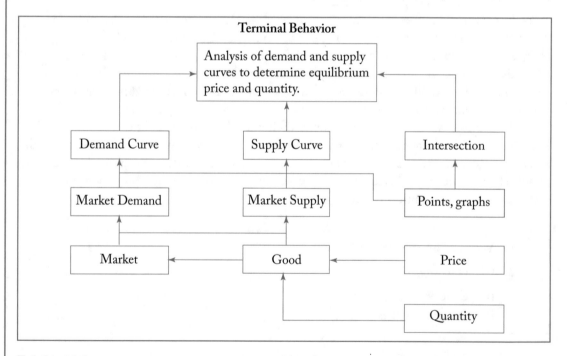

Exhibit 10.2

Both Ausubel and Gagné favored guided exposition (or guided learning) necessitating a sequential approach to instruction. Teachers need to be cognisant of the difficulties that many students have in learning concepts particularly complex concepts. Very often students get left behind because they have not understood a concept. This is particularly true of engineering,

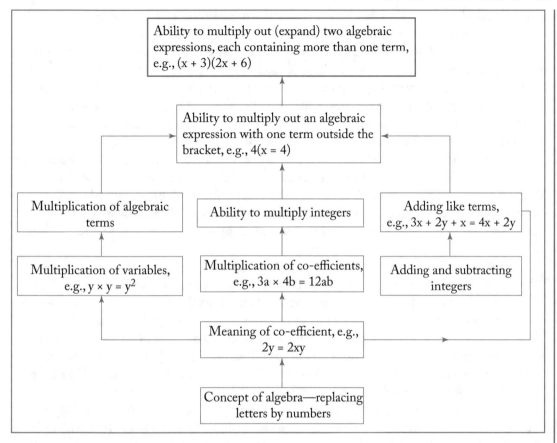

Exhibit 10.3

where it has also been found that motivation and course design enhance understanding in mechanics [3].

10.2 MISPERCEPTIONS

The difficulties that engineering students have in learning to think qualitatively were discussed in Journey 6. The need to understand concepts was highlighted. There is not only the possibility of not understanding but the possibility of misperceiving what is meant. In *Engineering Education* John Clement described a very similar problem to that described by Glyn Price (Exhibit 6.4) [4]. He gave detailed accounts of two interviews (verbal protocols) in which a student was asked to explain his problem solving processes. The first interview related to the concept of acceleration. It seemed that Jim the student had demonstrated an understanding of the concept because he had successfully obtained the acceleration of an object as a function of time. However, when

Jim was asked to draw a qualitative graph for the acceleration of a bicycle going through a valley between two hills he confounded the concept of acceleration with concepts of speed and distance. It appeared, wrote Clement, that while "Jim can use a symbol manipulation algorithm, his understanding of the underlying concept of acceleration is weak." The student has a procedure for getting the right answer in special cases but demonstrates little understanding of the concept when asked to apply it in the practical situation. We may describe such a student as having a "formula centered view of the concept." Other papers in engineering continue to draw attention to this phenomena as the following paragraphs will show.

At about the same time similar perceptions were being reported by scientists and the idea of "naïve" knowledge in science learning was promulgated [5]. Since then there have been hundreds of papers on misperceptions in the sciences and engineering. Bucciarelli takes an entirely positive view of the problem, he writes, "if misconceptions and common sense were somehow disallowed, we would still be living in the stone age. Popper is right: progress is the product of ill-conceived conjecture and its possible refutation" [6].

Accompanying them has been the development of "Concept Inventories," the first of which seems to have been in physics [7]. In their study of assessment in engineering, Pellegrino, DiBello and Brophy [8] consider concept inventories to have value in formative evaluation. This view is supported by an analysis of a number of inventories including the Concept Assessment Tools for Statics (CATS) developed by mechanical engineer Paul Steif [9]. Womeldorf has provided and introduction to the construction of these instruments [10]. As an alternative to concept inventories Kean et al have shown how the Model Eliciting Activities (MEA's) discussed chapter VI may be used to identify misconceptions [11].

Ruth Streveler and Ron Miller describe how the study of student understanding of concepts led to a change in course structure at the Colorado School of Mines. It is evident from that study, that in engineering, the educators have to differentiate between not-understanding, surface understanding, and deep understanding [12]. The primary problem is how to dislodge these misconceptions so that the new learning persists. Hsieh, Recktenwald and Edwards designed inquiry exercises to confront student misunderstandings but found that it was necessary prepare students familiar with traditional laboratory work for the inquiry based approach [13].

The more fundamental issue relates to "traditional" instruction since teaching must influence whether or not students misperceive concepts. I suspect that most beginning engineering educators are like my graduate student teachers and do not understand the importance of concepts in learning. Probably the most common technique used for teaching concepts and principles is to use examples. My graduate student teachers were surprised to find that the use of examples and non-examples during instruction is not as easy as it seems.

10.3   USING EXAMPLES

Engineering educator John Cowan wrote that conceptual understanding usually begins with examples. He had been convinced that this was the case by R. R. Skemp, a specialist in mathematics education. Skemp [14] believed, writes Cowan, "that it is essential that a concept is first encountered in the form of examples which establish the beginning of understanding. And he maintained that it is only when an initial understanding has been acquired, through the use and consideration of examples, that any abstract generalization or refinement of definition is possible or meaningful. For only at that point, he asserted has the learner developed sufficient understanding of the underlying concept on which to build thereon the theories and understanding which use and consolidate the concept" [15].

Cowan went on to describe how he had seen an elegant demonstration of this technique at an international conference during a keynote address on the acquisition of concepts.

The lecturer "taught her audience as she had taught her research subjects, the grammatical concept of the morpheme. First, she provided an assortment of examples, all of which were undoubtedly morphemes—and so this concept was established in the minds of her listeners—including me, who had not hitherto encountered it. Then she quickly tables a set of examples, all of which were not morphemes- although I might have a little earlier have so classified them, while I was still uncertain about what a morpheme is. Thus, the concept was yet more firmly concreted in the minds of the learners like me in the audience, as it had been in her research study. As her next step, and in refinement of our understanding, she gave us some more borderline examples which were marginally morphemes. By this point we had well and truly mastered the concept of the morpheme from examples" [16].

The strategy has some similarities with the early research that was done with young children on the sequencing of examples and non-examples. My graduate student teachers were asked to replicate one or other of these investigations. Many used a teaching strategy based on research that had been suggested by De Cecco and Crawford (Exhibit 10.4) [17].

1. Describe the performance expected of the student after the concept has been learned

2. Reduce the number of attributes to be learned in complex concepts and make important attributes dominant

3. Provide the student with useful verbal mediators

4. Provide positive and negative (non) examples in terms of appropriate number and realism

5. Present the examples in close succession or simultaneously

6. Provide occasions for student response and reinforcement of those responses

7. Assess the learning of the concept

Exhibit 10.4: de Cecco and Crawford's seven instructional steps for concept learning.

Steps 1 and 2 require the instructor to define the attributes and values of the concept (Exhibit 10.5). It is also helpful to young students to be able distinguish between conjunctive, disjunctive and relational concepts as it helps with the learning of grammar.

Attributes	Values
Conjugation	Given expression in personal pronouns
Changeability	Can be used as present/past participle
Tense	Can be singular or plural
Definition of behavior	Can express past, present, future
Linkage	Offer assertion
Regularity/irregularity within fixed range	Join words into sentence/phrase
	Change in conjugation

Exhibit 10.5: A graduate student teacher of modern languages view of the attributes and values of a "verb."

Mediators do not necessarily have to be verbal. They can be visual, as for example, concept cartoons which are attractive to students who do not like reading too much text [18]. But too much animation can be distracting [19]. It should be evident that computer assisted instruction can be designed to teach concepts in this way [20].

Often teaching is made difficult because the students bring with their thinking a stereotype of the concept (Exhibit 10.6). At any age the learner is likely to try and simplify the concept. Science concepts can be very difficult. One of my graduate student teachers argued that in teaching the concept of clouds to thirteen year olds difficult attributes such as structure, moisture content, and electrical charge should be ignored in favor of attributes such as height, shape and structure (see Exhibit 10.7).

Graduate student teachers find it difficult to define attributes and values. But, the evidence is, that when they are forced to think about the dominant features of a concept they find it to be an aid in the planning and implementation of a lesson. It is also evident that part of the confusion students have in learning concepts is because many teachers do not take a step-by-step approach that ensures the students understand the dominant attributes first. But this takes time and often tutors are unwilling to give that time because of beliefs about the need to cover the syllabus. This seems to be a central issue in teaching; it seems probable that a lot of the difficulties experienced by engineering students, especially in the freshmen year, arise from a shortage of time to assimilate the learning of the concepts being presented, especially when they are complex.

I had difficulty in convincing the students that "implicit costs" should be included as most of them has a stereotype that costs were comprised solely of "explicit costs" such as rent, wages, etc. Implicit costs include an item known as "normal profit," and the students were extremely reluctant to allow this to be classed as a cost of production. Until they overcame their original and slightly imperfect view of profit they were unable to appreciate the existence of these implicit costs. Thus concepts are highly influential as the student develops an understanding of his environment and all it entails. They can be important facilitators of education and instruction and allow generalizations to be made. However if the child enters the salesroom bearing a subjective idea of the concept that is not quite correct, then it can serve as an extremely harmful hindrance to effective learning. I found it very difficult to get them to appreciate that normal profit should be included. They had come armed with a stereotype that "normal profit" should be included. They had come armed with a stereotype that "profit" and "costs" should be entirely estranged.

Exhibit 10.6: Difficulties in the teaching of the concept "production costs" to sixteen year old boys experienced by a graduate student teacher.

NOTES AND REFERENCES

[1] Gagné, R. M. (1984). *The Conditions of Learning*, 4th ed., New York. Holt, Rinehart and Winston. 131

[2] *ibid* 132

[3] Vable, M. (2003). Enhancing understanding of concepts in mechanics of materials using design. *ASEE/IEEE Proceedings Frontiers in Education Conference*, 3, S3B-13 to 17. 133

In respect of the teaching a course in Material Mechanics, Vable concluded that the incorporation of design into the course was a powerful motivator and enhanced the understanding of generalized mechanics of materials concepts.

[4] Clement, J. (1981). Solving problems with formulas: Some limitations. *Engineering Education*, 25, pp. 150–162. 133

[5] Nasr, R., Hall, S. R., and P. Garick (2003). Student misconceptions in signals and systems and their origins. *ASEE/IEEE Proceedings Frontiers in Education Conference*, 1, T2E, pp. 23–28. 134

[6] Champagne, A. N., Gunstone, R. E., and J. C. Schwartz (1983). Naïve knowledge and science learning. *Research in Science and Technological Education*, 1(2), pp. 173–183. 134

[7] Bucciarelli, L. L. (2003). *Engineering Philosophy*. Delft, Netherlands. DUP Satellite. 134

"One thing that continues to surprise me each year, is the apparent inability of some of my students, and there are always some, to see the world—of bridges and buildings, forces and

A. Content	B. Method	C. Content	D. Method
Stage 1. Introduction			
Stimulation of interest in clouds	Introduce a discussion on rain. Its relevance to Irish climate. Is there any chance of predicting rain?	The use of non- examples	1. As the slides are shown a kettle is allowed to boil so that when the lights are turned on a layer of steam occurs in the upper levels of the classroom. 2. Breaking the laws on smoking in public places a cigar is then lit 3. A picture of Dublin smog is then passed around 4. The pupils are then asked to look at a picture in the textbook which shows industrial pollution.
Give a definition of clouds	A succinct and straight forward definition of the concept is given and written on the board.	Discussion of non- examples	A discussion is initiated to find out why these non- examples cannot be included in our concept.
Ascertain any previous knowledge and understanding of the terms to be used	One must use verbal questioning to find out what they know to see if they have any misconceptions and to see if they have understood the previously related classes including the terms, condensation, saturation, precipitation.	**Stage 3 Application**	
The concept of the hydrological cycle	Review verbally the concept of the hydrological cycle (dealt with in previous lesson) and hand out a copy to which they must assign labels	Diagrams are drawn in books	A copy of diagram (ii) is taken down in the copy books as a visual reminder of the height, shape and attributes of clouds

Exhibit 10.7: A graduate student teacher's lesson plan for clouds for thirteen year olds showing the use of examples and non-examples. (*Continues.*)

Stage 2 Presentation		Stage 4. Conclusion	
An explanation of the attributes of clouds	The discussion is initiated by a review of their previous observations on the attributes of height and color. This is reinforced by a discussion on other attributes such as formation and shape.	A summary and reinforcement programme is used.	The class reviewed by discussion with the essential information being highlighted by writing on the board.
Classification of clouds	Using a diagram shown on PP di indicate height (diagram (ii) shown permanently—a number of slides of cloud types are shown (examples)		

Exhibit 10.7: (*Continued.*) A graduate student teacher's lesson plan for clouds for thirteen year olds showing the use of examples and non-examples.

torque—as I do. At times, they come up with the most bizarre questions when challenged with certain problems and yet seem unable to accept and digest the explanations I provide. They have what faculty of schools of education responsible for the preparation of science teachers tag 'serious misconceptions'—a topic about which there is an ever growing body of literature in the scholarly journals in science education. The 'naïve science,' or 'common sense science,' or still more tolerant 'alternate world view' of youth are all to be rooted out, washed away, to make room for the way things really work, e.g., the true stories about planetary motion; equal and opposite internal forces; uniformly accelerated motion and the like."

"I myself find this deviant behaviour on the part of my students refreshing and provocative. When so challenged I want to know where in the world they got this strange way of seeing things. What do they call upon to justify their misconceptions? If I can reconstruct something of the student's conceptual scheming—which he or she might or might not acknowledge as their way of thinking—then I have a much better chance of success at conversion of the student from error to my way of seeing."

"Misconceptions are not necessarily disabling. Common sense ordinarily serves us well as a basis for thinking, acting and social exchange (until its undoing by science). Unquestioned presumptions on the one hand and long dead myth and metaphor on the other, are normally harmless. In fact in ordinary times, they are enabling……" p. 5.

[8] Hestenes, D., Wells, M., and G. Swackhamer (1992). Force Concept Inventory. The *Physics Teacher*, 30, pp. 159–166. 134

[9] Pellegrino, J. W., DiBello, L. V., and S. P. Brophy (2016). The science and design of assessment in engineering in A. Johri and B. M. Olds (Eds.), *Cambridge Handbook in Engineering Education Research*. New York Cambridge University Press. 134

[10] Steif, P. S. and M. A. Hansen (2007). New practices for administering and analysing the results of concept inventories. *Journal of Engineering Education*, 96, pp. 205–212. 134

[11] Womeldorf, C. (2007). An introduction to the construction of engineering concept inventories: Tools for impacting teaching, learning and assessment. *Proceedings of the Spring 2007 American Society for Engineering Education North Central Section Conference*. West Virginia Institute of technology. March 30–31. 134

[12] Kean, A. et al. (2017). Identifying robust student misconceptions in thermal science using model-eliciting activities. *Proceedings Annual Conference of the American Society for Engineering Education*. Paper 2916. 134

[13] Streveler R. A. and R. L. Miller (2001). Investigating student misperceptions in the design process using multidimensional scaling, *ASEE/IEEE Proceedings Annual Conference American Society of Engineering Education*, Paper 2630. 134

The instructor identified 32 concepts (or terms) which the students were asked to cluster. This exercise was repeated again during the last week of instruction. In this way changes resulting from the course could be evaluated and areas of difficulty identified. Four clusters relating to economic analysis, energy transfer, analysis of processes, and heuristics were revealed. Some of the terms were scattered around the clusters. The post-test found some reshuffling. The four clusters remained but there were some additions to the third cluster. The heuristic terms became more closely structured; however, the terms are not, in practice, related in any fundamental way. This finding led the investigators, to suggest that the students might not have a deep understanding of these terms. Similarly, the term "life cycle analysis" did not seem to be understood, and it seemed that students might not have an understanding of how trouble shooting related to process design and analysis. This evaluation led to changes in course structure.

[14] Hsieh, C., Recktenwald, G., and R. Edwards (2017). Implementing inquiry based experiments in a fluid science laboratory. *Proceedings Annual Conference of the American Society for Engineering Education*. Paper 1351. 135

[15] Skemp, R. R. (1971). *The Psychology of Learning Mathematics*. Harmondsworth. Penguin. 135

[16] Cowan, J. (2006). *On Becoming an Innovative University Teacher. Reflection in Action.* 2nd ed., Maidenhead. SRHE/Open University Books. 135

[17] *ibid* 135

[18] de Cecco, J. P. and W. R. Crawford (1974). *The Psychology of Learning and Instruction.* Englewood Cliffs, NJ, Prentice-Hall. 136

[19] Keogh, B., Naylor, S., and C. Wilson (1998). Concept cartoons. A new perspective on physics education. *Physics Education*, 3394, pp. 219–225. 136

They reported that "too much animation can be distracting." There was too much "flash." This study gives some support to the view that that the same rules that govern the learning of concepts apply in computer assisted learning environments. In favour of animation, it has been pointed out that where the behavior of a physical phenomenon changes with changes in the parameters, animation is an ideal way of learning the concept. Khaliq described how in a course on integrated circuit fabrication multimedia modules were used to show the fabrication of the device (i).

Another investigation at RMIT Melbourne showed how animations could be used to help students discover their misconceptions. The investigators designed a simulated laboratory setting that enabled the students to specify and test their own expectations of heat transfer in an experimental situation. At the same time it enabled the investigators to better understand the students' difficulties. "We wanted the students to predict what they expected to see, so that they would attend with more interest to their observations and hopefully identify and puzzle over any discrepancies. To facilitate this, we asked the students to work in pairs so that they could discuss their ideas as they evolved. The 'explain' phase of the task required students to reflect on the surprises they have encountered and to develop and refine hypotheses, rather than restate a theoretical position which they had already been taught" (ii). It is reported that the approach worked well, and it was noted that students' had more difficulty with unsteady state conditions than with steady state ones.

Segall developed a new freshmen level course that illustrated basic engineering concepts and principles by means of science fiction films and literature. "Central to the course delivery is 'poking' fun at the disobedience of the laws of nature and the misuse of engineering while at the same time teaching the correct behaviours." Part of the assessment required the students to describe and explain at least five events where they believed the laws of physics were observed/and or violated. They were also asked to discuss any technology/society/ethical issues raised by the story (iii).

(i) Khaliq, M. A. (2001). Interactive multimedia courseware for integrated circuit fabrication course. *ASEE/IEEE Proceedings Frontiers in Education Conference*, 3, S1C-1 to 4.

(ii) Ball, J. and K. Patrick (1999). Learning about heat transfer. "Oh I See" experiences *ASEE/IEEE Proceedings Frontiers in Education Conference*, 2, 12e5-1 to 6.

(iii) Segall, A. E. (2002). Science fiction in the engineering classroom to help teach basic concepts and promote the profession. *Journal of Engineering Education*, 91(4), pp. 419–424.

[20] Davidovic, A., Warren, J., and E. Trichina (2003). Learning benefits of structural example-based adaptive tutoring systems. *IEEE Transactions on Education*, 46(92), pp. 241–251. 136

The screen of their generic computer based instruction system showed a table of contents, an introduction box that introduced and defined new concepts to be learned, an explanation, and a box in which the concepts were explained in greater detail. The main feature was the provision of two boxes side-by-side for presentation of two examples that could be compared. The purpose was to provide multiple examples and through them enable the student to gain a deeper understanding that would lead to better and deeper concept maps. Below the two boxes were to other boxes for student work. An initial evaluation with 117 students in a one-hour tutorial suggested that the rate and extent of learning was significantly greater than when the features were used alone or both were absent.

JOURNEY   11

Complex Concepts

11.1 COMPLEX AND FUZZY CONCEPTS

Students sometimes have difficulty in understanding complex concepts because they are looking for black and white definitions, the effect of which is to restrict their understanding. "If you initially conceptualize an issue in an over restrictive way, this can prevent later insights from developing, committing you to a single track of thinking" [1]. Freshmen students in the social sciences and humanities are often faced with extremely difficult concepts, as for example "democracy."

Dunleavy recommended the five step approach to solving complex problems shown in Exhibit 11.1. Step 1 is completed by placing democracy in the universe of the political system. To complete step 2 the student has to ask, "What is the opposite of democracy?" Some of the dichotomies proposed will be appropriate and others inappropriate. Dunleavy shows how easy it is to produce an inappropriate idea. For example, "the familiar contrast between 'democracy' and 'totalitarianism' is a false dichotomy because totalitarian regimes are a very small sub-class of non-democratic regimes…but most non-democratic regimes do not go this far" [2]. False dichotomies can arise from or be stereotypes.

---

1. Place the concept in its universe.

2. Search acronyms to the concept within this universe.

3. Look for antonyms or potential antonyms.

4. Look for unstated partner words.

5. Explicitly examine different forms of the concept.

Exhibit 11.1: Dunleavy's recommended five-step approach to the clarification of concepts.

Similar complexities will be met in science. Howard makes the point that some stimuli are hard to classify as "exemplars." He cites the example of a teacher who used the example of "species" in a course on evolution only to pull the notion itself apart at the end of the course [3]. Miller and his colleagues at the Colorado School of Mines obtained the views of teachers in a variety of engineering topics about the difficulty of different concepts in their subjects, and concluded that, as in the humanities, there were concepts in engineering that were fuzzy, that is, having no clear cut boundaries or defining features [4].

A useful procedure for teaching complex concepts is that of a "best example" or "proto-type," the latter being the most typical case among a category of members. A stimulus identifies a category as a function of its resemblance to a prototype [5]. One approach is to define the concept, present the learners with one or two typical exemplars with the instruction to remember them by forming a visual image so as to memorize their features. The learners are then presented with a series of examples and non-examples among which is the best example. They categorize the examples and non-examples by reference to the best example (or prototype) and in so doing learn how to generalize and discriminate against the prototype. They can also be asked to define the dimensions along which the examples vary from the best example [6].

Analogy and metaphor are also commonly used in the teaching of easy and complex concepts. Analogies with water systems are often used by teachers to explain electrical circuits.

Gordon (cited by Howard) distinguishes between three types of metaphor, direct analogy, personal analogy and compressed conflict. The first involves comparison of two concepts (X is like Y). The second asks the learner to put themselves in the position of someone or thing. "What would it be like if I were…?" The third is the comparison of two contradictory concepts [7].

Metaphors can cause confusion and misunderstanding. There is a danger that if incorrect features are transferred considerable misunderstanding will ensue. For this reason it is important that students are familiar with one of the domains. It is, therefore, important to check that the metaphor has been understood.

When they were writing their final evaluations on concept learning my graduate student teachers were told to read certain chapters from a book by R. W. Howard. They were then asked to say if they would have used a different approach to teaching the concept, and say what that approach would be. Most were happy with what they had done, but their favorite alternative was the concept or semantic map.

11.2   STAGED DEVELOPMENT

There has long been a debate about the value of teaching real life applications in science and technology as a means of helping student understanding. Raghavan and his colleagues consider that such an approach helps students better understand mathematical concepts. One of them had developed modules designed to increase complexity incrementally. For example, "a module in spring-mass systems started with a linear spring system, transited to a non-linear spring model, and finally ended with a coupled spring system model for the landing system of a spacecraft" [8].

Unfortunately, Raghavan and his colleagues had not tried the model out on their students but they describe in detail how the modules are constructed based on the instructional strategy of guided inquiry. Such schemes might benefit from a spiral approach.

11.3   CONCEPT MAPPING AND KEY CONCEPTS

The role of concept maps in distinguishing experts from novices and the implications for teaching have already been discussed in Section 5.4. They are techniques for facilitating meaningful learning [9] by organizing the essential information into a visual framework that displays the attributes and values of the concept to be learned. They may be used to design and evaluate instruction, to diagnose what students know, and act as advanced organizers. In their turn students can use them to learn in lectures, and plan their learning. Such maps come in all shapes and sizes.

In the Anglo-Irish system of examining where undergraduates write essays or solve problems that take from 40 minutes to 1 hour, students are encouraged to prepare a summary before they write their answers. Exhibit 11.2 shows a concept map that one of my graduate student teachers drew before answering a 1 hour question on self-accountability in teaching [10]. The schematics illustrating Gagné's approach to instruction shown in Journey 10 are concept maps. Their value in learning is enhanced if students are helped to use them to construct frames of reference that will help them deal with new situations.

Krause and Tasooji found that if concept sketching was used to implement pair discussions that a high gain in understanding in the Materials Concept Inventory was attained. It appeared to enhance learning more than paired discussions. Their results suggested that "concept sketching may facilitate repair of students" conceptual frameworks through the displacement of the robust misconceptions held by the students [11]. Like other engineering educators they found the National Academies report *How People Learn: Bridging Research and Practice* to be a valuable support [12]. One study supplemented the framework presented in this study with the Legacy Learning cycle for a challenge in solid mechanics [13].

A curriculum may be designed around concepts, some of which are more central than others. These latter have been called key concepts. They are procedural devices that have as their purpose the design of the structure and content of the course. Taba who is credited with the invention of the idea uses difference, multiple, causation, interdependence and democracy as examples. Taba writes, "These types of concepts are usually in the background and therefore often relegated to incidental teaching. In a sound curriculum development they should constitute (what some have called) recurrent themes, the threads which run throughout the (entire) curriculum in a cumulative overarching pattern" [14]. An example is given in Exhibit 11.3. The relevance of the spiral curriculum may be appreciated from this list which was used to develop a curriculum for middle schools in England by W. A. L. Blyth and his colleagues [15].

The purpose of key concepts is to help the instructor choose, and organize topics for work (e.g., Exhibit 11.4). The interpretation of a syllabus (content) in terms of its concepts is an essential task, for it helps teachers concentrate on those tasks necessary for understanding, and the transfer of learning. Key concepts are therefore, objectives to be achieved and are as important as behavioral objectives. Moreover, given the understanding that often instruction is

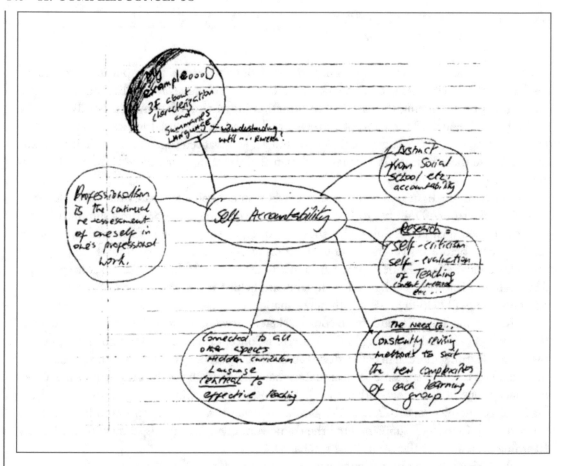

Exhibit 11.2: Concept map drawn by a graduate student teacher prior to writing a 1 hour answer to a question on self-accountability in education in a university examination.

carried out at too fast a rate, a pace that links concepts to objectives should ensure a focused and coherent curriculum that will bring about high order thinking.

Technology based concept mapping has been introduced as an active learning strategy to help both students and instructors create visual navigation structures through complex knowledge domains such as the content of a course or a curriculum [16].

Related to the idea of the key concept is the idea of "concept clusters." Engineering educator Paul Steif has shown how he has used "concept clusters" to organize both assessment and instruction in Statics [17]. He distinguishes between (a) skills that are actions that can be mastered by rote practice, and (b) concepts that demand much more careful explanation and deeper understanding (Exhibit 11.5). He argues that some errors may stem from inadequate skills rather

1.	Communication	The significant movement of individuals groups or resources or, the transmission of significant information.
2.	Power	**The purposive exercise of power over individuals and society's resources.**
3.	Values and beliefs	The conscious or unconscious systems by which individuals and societies organize their response to natural social and supernatural disorders.
4.	Conflict/consensus	**The ways in which individuals and groups adjust their behavior to natural and social circumstances.**
5.	Similarity/difference	Classification of phenomena according to relevant criteria.
6.	Continuity/change	**Distinction of phenomena along this essentially historical dimension.**
7.	Causes and consequences	The notion that change in a state of affairs can be contributed to the phenomena preceding.

Exhibit 11.3: Key concepts in the University of Liverpool's History, Geography, and Social Sciences Middle schools project.

than conceptual misunderstandings. Related to this is an argument by Papadopoulos that procedural knowledge should also be emphasized if there is to be conceptual understanding [18].

Baillie, Goodhew and Skryabina have explored the use of threshold concepts in engineering for removing potential blocks in student understanding [19]. Threshold concepts were suggested by economists Meyer and Land [20]. They are certainly complex concepts and seem to fulfil the same functions as key concepts. In-depth understanding is likely to produce an "ah-ha" moment and to have in Meyer and Land's terminology a transformative effect. Similarly, a key concept once learned is likely to be built into one's schema and not forgotten. Moreover, it is necessarily part of a more general map whose understanding will depend on similar understandings of all the concepts in the map. Examples of threshold concepts in the literature include standard deviation and opportunity cost. The concept of waves would fall into this category, and Donald's key concept structure for waves serves to affirm this argument (Exhibit 11.6).

In the United States H. Lynn Erickson has questioned the standards movement's ability to raise the level of conceptual thinking, and argues that national standards have to be looked at through concepts so that thinking can be taken beyond facts, and facilitate understanding. Given the arguments about the crowded curriculum it is surprising that engineering educators have not started to examine the curriculum from the combined focus of outcomes and key concepts.

Objectives	Key Concepts	Specific Content	Learning Experience
Stage 1. Introduction			
1. The fostering of willingness to explore personal attitudes and values and to relate these in other people. 2. To understand how other people interact with their particular environments. 3. To make connections between concepts and percepts which have been learned in previous lessons to the present analyzes. 4. The encouraging of an openness to the possibility of change in attitudes and values.	**ENVIRONMENTS** (similarity and difference)	1. Tropical lands 2. Temperate lands 3. Cold lands	Group work. Various groups make a study of these three environments. Work could be divided into (a) physical environment (b) social environment (c) cultural environment. Overview
1. To develop in the students an ability to plan. 2. To apply theories to new situations and to evaluate the result. 3. To develop and test hypotheses. 4. To work in a group and to coordinate efforts and delegate tasks.	**URBANIZATION** (continuity and change)	1. Urbanization on a global scale. (a) Process over time (b) Third world cities (problems and growth) 2. Economic development (a) Agriculture to industry. (b) Stages of development and the third world, (c) Theories-Growth centers vs. decentralization 3. Planning city growth (a) Urban sprawl and decay. (b) Renovation and renewal (c) Rehousing and new town. 4. Shannon development 5. Dublin	Application of Rostow's stages of development. Evaluation of theories Group project 1. Fieldwork around Dublin collecting information. 2. An urban study of Dublin emphasizing problems associated with growth and proposing possible solutions.

Exhibit 11.4: Extract from a curriculum in geography developed by Gina Plunkett a graduate student teacher. Part of a third year curriculum (reprinted in Heywood, J. (1982)). Pitfalls and Planning in Student Teaching. Kogan Page. London.

Concept Clusters

1. Forces acing between bodies.

2. Combination and/or distribution of forces acting on a body are statically to a force and a couple.

3. Conditions of contact between bodies or types of bodies imply simplification of forces.

4. Equilibrium conditions are imposed on a body.

The skills needed for implementing the concepts of Statics are

1. Discern separate parts of an assembly and where each connects with the others.

2. Discern the surfaces of contact between connected parts and/or the relative motions that are permitted between two connected parts.

3. Group separate parts of an assembly in various ways and discern external parts that contact a chosen group.

4. Translate the forces and couples which could be exerted as a connection (e.g. there is only a force in one direction) into the variables, constants and vectors that represent them.

Exhibit 11.5: Paul Steif's concept clusters and skills for teaching and learning Statics.

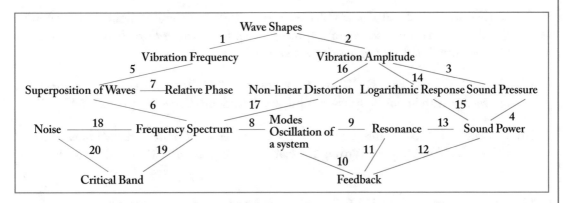

Exhibit 11.6: Donald's key concept structure of "Waves" [21].

There is quite a body of research in engineering education that shows the importance of concept learning in engineering.

NOTES AND REFERENCES

[1] Dunleavy, J. P. (1986). *Studying for a Degree in the Humanities and Social Science*. London. Macmillan education. 143

[2] *ibid* 143

[3] Howard, R. W. (1989). *Concepts and Schemata. An Introduction*. London, Cassell. 143

[4] Miller, R. L., Streveler, R. A., Olds, B. M., and M. A. Nelson (2004). Concept based engineering education: Designing instruction to facilitate student understanding of difficult concepts in science and engineering. *ASEE/IEEE Proceedings Frontiers in Education Conference*, SA1, 1 to 2. 144

[5] Howard cites Rosch, E. (1973). On the internal structure of perceptual and semantic categories in T. E. Moor (Ed.), *Cognitive Development and the Acquisition of Language*. New York. Academic Press. 144

[6] *ibid*. As described by Howard from Tennyson, R. D. and O. Park (1980). The teaching of concepts. A review of instructional design literature. *Review of Educational Research*, 50, pp. 55–70. See also Tennyson, R. D. and M. J. Cocchiarella (1986). An empirically based instructional design theory for teaching concepts. *Review Educational Research*, 56, pp. 40–71. 144

[7] *ibid* 144

[8] Raghavan, J. et al. (2017). Problem-solving experience through light-dose computational mathematical modules for engineering students. *Proceedings Annual Conference of the American Society for Engineering Education*. Paper 2721. 144

[9] Novak, J. D., Godwin, D. B., and G. Johnson (1983). The use of concept mapping in knowledge with junior high school students. *Science Education*, 67, pp. 625–645. 145

[10] Heywood, J. (2008). *Instructional and Curriculum Leadership. Towards Inquiry Based Schools*. Dublin. Original Writing for National Association of Principals and Deputies. 145

[11] Krause, S. and A. Tasooji (2017). An intervention using concept sketching for addressing dislocation-related misconceptions in introductory materials classes. *Proceedings Annual Conference American Society for Engineering Education*. Paper 2026. 145

[12] Donovan, M. S., Bransford, J. D., and J. W. Pellegrino (Eds.), (1999). *How People Learn. Bridging Research and Practice*. Washington, DC, National Academies Press. 145

[13] Fuentes, A., Crown, S., and R. Freeman (2017). Human bone solid mechanics challenge: Functionally graded material structure with complex geometry and loading. *Proceedings Annual Conference American Society for Engineering Education*. Paper 2056. 145

[14] Taba, H. (1962). *Curriculum Development. Theory and Practice*. New York. Harcourt Brace. 145

The original quotation uses "centre" not "entire." The problem arises from the use of the term curriculum which may relate to a single subject or to all the subjects that make up an area of study, e.g., engineering or in contrast liberal-arts curriculum.

[15] Blyth, W. A. L. et al. (1972). *Pace, Time and Society. An Introduction*. Bristol, Collins, pp. 8–13. 145

[16] Miertschein, S. and C. Willis (2017). Using course maps to enhance navigation of E-Learning environment. *Proceedings Annual Conference of the American Society for Engineering Education*. Paper 2363. 146

[17] Steif, P. (2004). An articulation of the concepts and skills which underlie engineering statics. *ASEE/IEEE Proceedings Frontiers in Education Conference*, F1F, pp. 5–10. 146

[18] Papadopoulos, C. (2017). Assessing cognitive reasoning and learning in mechanics. *Proceedings Annual Conference of the American Society for Engineering Education*. Paper 2537. 147

[19] Erickson, H. Lynn (1998). *Concept-based Curriculum and Instruction. Teaching Beyond the Facts*. Thousand Oaks, CA, Corwin. 147

"Does raising 'standards' mean learning more content, which is delineated through 'objectives'? Or does it mean using critical content as a tool to understanding key concepts and principles of a discipline, and applying understanding in the context of a complex performance? From a review of national standards, it is clear that most disciplines favour the latter goal. Certainly knowing (and often memorizing) a body of critical content knowledge is important for an educated person. But conventional models of curriculum design have focused so heavily on the information level that most teachers lack training for teaching beyond the facts. Yet the standards and newer assessments assume that students will demonstrate complex thinking, deeper understanding, and sophisticated performance."

[20] Baillie, C., Goodhew, P., and E. Skryabina (2006). Threshold concepts in engineering education-exploring potentail blocks in student understanding. *International Journal of Engineering Education*, 22(5), 955–962. 147

[21] Donald, J.G. (1982). Knowledge structure: methods for explaning course content. *Journal of Higher Education*, 54(1), 31–34. 149

JOURNEY 12

The Learning Centered Ideology–How Much Should We Know About Our Students?

12.1 INTRODUCTION

The Learning Centered ideology is in stark contrast to the social efficiency ideology. The child is at the center of, and has a profound influence on the curriculum process. This ideology is associated with the educational philosophy of John Dewey. It holds that the learning centered school should be totally different to the traditional school, and more like the Montessori and Reggio Emilia schools. They are activity based.

Learner Centered schools are based on natural developmental growth rather than on demands external to it. "Individuals grow and learn intellectually, socially, emotionally and physically in their own unique and idiosyncratic ways and at their own individual rates rather than at a uniform manner" [1, p. 111]. In the UK, the learner centered approach was espoused in the Plowden report [2].

The philosophy that underpins these schools is constructivism. The schools and curriculum are designed to produce students who are "self-activated makers of meaning, as actively self-propelled agents of their own growth, and not as passive organisms to be filled or moulded by agents outside themselves" [1, p. 115]. Learning moves from the concrete to the abstract.

The idea of active and passive learning has become part of the vocabulary of higher education, not in the sense of organizing an institution for active learning, but in the sense of teachers organizing and managing their classrooms for student centered active learning. The relationship between the teacher and the student is quite different to that established by educators from either the scholar academic or social efficiency ideologies, and Cowan [3] argues, to be preferred.

Because knowledge is created by individuals as they interact with their environment, the objectives of a learner centered education are statements of the "experiences" the student should have. This view brings learner centered educators into conflict with those who believe that the objectives of an education are its measurable outcomes, which is the case with ABET and other systems where politicians require measures of efficiency.

In addition to establishing an environment for learning the teacher has the functions of observing and diagnosing individual needs and interests together with facilitating the growth of the students in their care.

Learning centered educators are opposed to the psychometric view of testing as expressed by social efficiency educators. Standardized tests are anathema to learner centered educators. It is assumed that children's work is best assessed by children themselves hence the importance of learning logs and journals. Some engineering educators are advocates of peer and self-assessment as well as the use of portfolios [4, 5, 6, 7]. Learning Centered tutors create communities of engagement, communities that care.

12.2   COMMUNITIES OF PRACTICE, COMMUNITIES THAT CARE

Smith, Smith and Felder described a learner centered one week course for eleven rising sophister female students at Smith College who were at risk in physics and some mathematics skills [8]. They took the view that learners construct knowledge only after they have encountered and used knowledge in a social context (social cognitive theory). Thus, this course established a "social context that supported collaboration, shared thinking, and risk taking, and that facilitated various kinds of engagement with the content in ways likely to build understanding."

Part of day one was devoted to peer teaching "when student teams took turns demonstrating and explaining complicated potion-time and velocity-time graphs to the rest of the class." Peer teaching and peer review (assessment) [9] have a very long history in educational practice, and recent research has confirmed the benefits they bring to learning [10]. On the second day the students "decided to spend part of day finishing plotting activities from Day 1." In another activity on the second day they were given engineering exam questions and checked sample answers by evaluating the units and limits of the answer. "The instructor continued to model the framework by using it to evaluate answers throughout the week." This was based on the Expert Problem Solving Framework. On the third day which focused on Newton's laws "students completed a discussion-activity session that focused on addressing misconceptions related to friction forces felt while walking and running-were used to direct discussion." Kinesthetic activities played an important role in the provision of hands-on activities. A recent study obtained promising results from the use if kinesthetcs in helping learners understand some basic principles of physics [11].

From the course description the tutors certainly acted as facilitators for a scheme that they had designed, but within that scheme students played a major role in determining its direction. They began to become a community of practice [12]. But there is surely a need to become a community of care [13].

Many of my beginning teachers were enthused by the teaching experiments they had to do but this enthusiasm was often lost because the culture of the staffroom was built around "chalk and talk" with a view to getting students through public examinations the questions for which could be predicted and the answers memorised. Similarly, as I reported years ago managers who were enthused by ideas they had learned on in-company training courses became depressed when no one listed to their ideas when they returned to work [14]. It is essential that such courses are followed up especially in the first year of degree programmes.

Peters and Pears remind us that in the first year students in computer science often have low self-efficacy, a point alluded to in several studies of engineering students [15]. Women, in particular, often perceive themselves as being less capable than males. More significantly the first year of a programme is when students examine and search for their identity while negotiating meaning. How they engage with the course is therefore important. Their investigation into how students participated in an introductory computer studies course leads to the view that courses in general need to examine how students participate in them and the extent to which they provide experiences for participation [16]. It follows that for the full benefits of a course like Smith, Smith and Felder described are to be realized that the same approach would have to be continued into their main studies.

It is depressing to think that changes of this kind are often opposed because students have been led to believe through their prior educational experience that they are not likely to learn as well as they would in traditional settings in spite of research to the contrary [17] (private communication from Mani Mina).

Smith, Smith and Felder's course certainly humanized this engineering education experience just as Schiro argues the learner centered approach did to school education.

This ideology promoted the "view that teachers are facilitators of learning, [the] introduction of personal meaning to our vocabulary as a means of speaking about knowledge, [and] bringing to our attention the importance of different learning styles, [the] integration of the curricula through the use of projects......" [1, p. 148]. This chapter is concerned with what we can learn about students from their learning styles and temperament. Understanding student differences is an obligation of every teacher.

How much do we need to know about them? Engineering educators have demonstrated that a valuable dimension of student behavior is their learning style.

12.3 LEARNING STYLES

We all have preferred ways of organizing what we see and think about, or—different styles of conceptualization and patterning of activities. Styles are dispositions that students bring with them to their learning. Strategies are approaches they learn as a result of their attempts to adapt to the learning environment more especially the assessment tasks that are set. Learning styles have been shown to be related to learning and studying, and the teaching styles of the instructor.

Numerous learning styles have been described. Anthony Grasha has found that the factors that contribute to learning styles can be grouped together in the five categories shown in Exhibit 12.1.

Cognitive	Relating to the acquisition, retention and retrieval of information.
Sensory	Relating to the acquisition of information via the senses.
Interpersonal	Relating to the acquisition of information within social groupings and groups, influenced therefore by roles and role expectations, group norms, leadership and discourse (occasions of formal and informal learning).
Intrapersonal	Relating to the influence of the individual on him/herself. Needs and motives and especially the thinking needed for self-control.
Environmental	Relating to the physical environment in which we learn and the resources provided.

Exhibit 12.1: Grasha's categorization of the factors said to contribute to learning styles (Grasha, A. (1984)). The journey from Greenwich observatory (1796) to the college classroom. *Improving College and University Teaching*, 32, (1), pp. 46–53.

Although this chapter is primarily concerned with the use of David Kolb's learning styles in engineering, and the work of engineering educator Rich Felder to develop an Index of Learning Styles [18], there are many other styles that may be considered [19].

12.4 CONVERGENT AND DIVERGENT THINKING

Convergent and divergent thinking styles are possibly the best known thinking styles. Although not a direct measure of creativity divergent styles and tests of divergence are considered to give some indication of the potential that a person has to be creative. The balance between convergence and divergence has been found to be a good indicator of performance among a group of electrical engineering students in the UK [20]. Guilford considered that effective thinking resulted from the sequential use of convergent and divergent processes a point that was illustrated by engineer psychologist P. R. Whitfield whose analysis of the engineering problem solving process is shown in Exhibit 12.2 [21].

Although it is not part of the purpose of this chapter to pursue a study of creativity it should be noted that in schools it has been found that much teaching and assessment encourages convergent thinking. This is consistent with the findings of Perry on intellectual development in higher education, and not without significance for engineering educators.

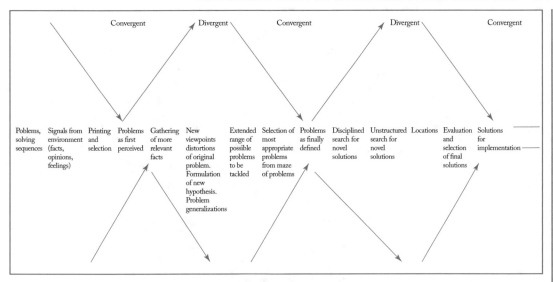

Exhibit 12.2: Part of Whitfield's illustration of the innovative (problem-solving) process (abbreviated—selected to show the significance of convergent and divergent thinking in the process). In the original diagram the phases were paralleled with sections for desirable personality characteristics, supporting techniques and personal development methods. Whitfield, P. R. (1975). *Creativity in Industry*. Harmondsworth. Pengui.

12.5 KOLB'S THEORY OF EXPERIENTIAL LEARNING

Kolb's theory of experiential learning is in the tradition of Piaget [22]. The learning of concepts involves four processes. First, comes a specific experience that causes the learner to want to know more about that experience (concrete experience). For that to happen the learner has to reflect on that experience from as many viewpoints as possible (reflective observation). From this reflection the learner draws conclusions (abstract conceptualization) and uses them to influence decision making or action (active documentation). The cycle draws the learner into a form of reflective practice. The axes (Exhibit 12.3) represent the available information of abstraction contained in the experience (y-axis) and, the processing of information through reflection or action on the conclusions drawn (Y-axes).

Kolb proposes that we each have dispositions to think in the style associated with one of these activities, and that in any group of people one is likely to find persons with different learning styles. Further, learning styles will differentiate groups of particular professionals from each other. The implications for teaching are profound. Thus, a teacher who wishes to teach a concept or principal should teach it in four different ways even though he/she has a preference for one style. This means they will have to be cognisant of a whole range of instructional strategies.

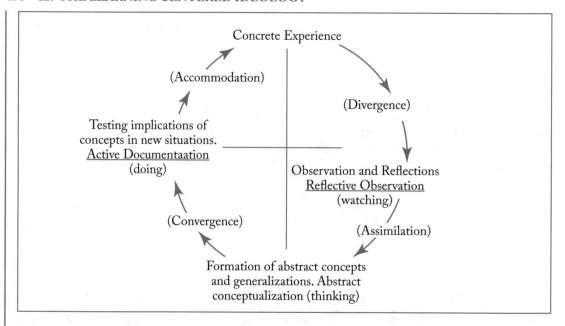

Exhibit 12.3: An adaption of Kolb's experiential learning model based on FizGibbon, A. (1987). Kolb's experiential learning model as a model for supervision of classroom teaching for student teachers. *European Journal of Teacher Education*, 10(2), pp. 163–178, and Stice, J. E. (1987). Using Kolb's cycle to improve student learning. *Engineering Education*, 77(5), pp. 291–296.

Necessarily the learners would have to learn in each of the styles. Kolb designed "The Learning Styles Inventory" to measure the preferred learning styles of individuals.

I asked my graduate student teachers to test this model in their classes, and to answer the question, "Should learning styles be matched to teaching styles?" Many found that their teaching style was appropriate for some pupils but not for others, and that they had to change their teaching from whole class instruction based on their style alone [23]. They found that designing test questions to match each style difficult. A lesson plan in mathematics that one of them undertook is shown in Exhibit 12.4.

Following my colleague Anne FitzGibbon's explanation, [24] the four Kolb learning styles are: (1) Divergers who like to "imagine" and generate ideas. They are emotional and relate well to other people. They do not perform well in tests that demand single solutions. Divergence relates to that part of the problem-solving process that identifies differences (problems) and compares them with reality (first quadrant). (2) Assimilators (second quadrant) are concerned with abstract concepts. They are interested in the precise and logical development of theory rather than with its application. Assimilation relates to the solution of problems, and the consideration of alternative solutions in the problem solving process. (3) Convergers (third quadrant) are the

Lesson Phases	Content	Learning Strategies	Imagery
Introduction	Explain to students about their being four types of learning style and that today's class is investigating that theory.		
Concrete experience **Overall aims:** (1) to extend students understanding of area (2) to introduce four different learning and teaching styles **Non-behavioral objectives:** (1) to introduce the concept of Simpson's rule (2) to illustrate the use of guided imagery in mathematics. **Behavioral objctives:** By the end of the lesson the students will be able to: (1) describe in their own words, what is meant by area. (2) state the purpose of Simpson's rule (3) give examples of where Simpson's rule can be applied (4) apply Simpson's rule mathematically.	Guided cognitive imagery. Students take part in this exercise on shape: the area of shape with irregular boundaries (see column four)	Students participate individually. Speaking to entire class	Initially 5 minutes was taken to get them to sit back and relax using the method involving awareness of surroundings and self. *"I then invited them to imagine an uneven shaped field/piece of land/ island (or whatever came to mind). I then asked them questions about its color, scent, were there any flowers? Animals? They were then invited to get into a helicopter and fly all along the outside/periphery of the field and to examine what was around it,- was it fenced off, an island etc. The students were asked to picture the boundaries of the field and to imagine flying the exact path of the boundary-twisting and turning along, inspecting every section of the boundary."* Time was allowed throughout to allow for thought and imagination.

Exhibit 12.4: A mathematics lesson for 15–16 year old students (girls) by a graduate student teacher. (Edited to show imagery exercise and objectives). (From Heywood, J. (2008). *Instructional and Curriculum Leadership. Toward Inquiry Oriented Schools.* Dublin, Original Writing for National Association of Principals and Deputies.). (*Continues.*)

Reflective observation	Students asked to write down what shapes they saw etc as a result of the imagery exercise. Discuss in groups of 2/3 what each person experienced-. Discuss in same groups of why they think, in everyday life, it would be necessary to know the area of such shapes, and where it would be useful. Group opinions are written down on an OHP	Accommodators and divergers should benefit from group work. Converger's should gain from written exercise.
Abstract conceptualization	Draw shape with irregular boundary on board-divide into strips of equal width and assign y, etc to lengths-give formula for Simpson's rule-explain basis of formula-Give easy method for remembering formula-sample problems solved on board and then by individuals	Students work individually and are given time to ask questions. Should appeal to convergers.
Active experimentation	Students asked to recall the shape they pictured in the imagery exercise. Each asked to make that shape with a piece of cardboard-Students asked to predict area of the shape it most resembles-then divide them into strips and apply Simpson's rule—compare the result with the estimate (ensure all using correct units of measurement)	Students work individually on making own model and applying Simpson's rule. Expect assimilators to gain most.
Conclusion	Reinforcement of lesson using questions to test behavioral objectives. Collect written descriptions and cardboards shapes.	Questioning.

Exhibit 12.4: (*Continued.*) A mathematics lesson for 15–16 year old students (girls) by a graduate student teacher. (Edited to show imagery exercise and objectives). (From Heywood, J. (2008). *Instructional and Curriculum Leadership. Toward Inquiry Oriented Schools.* Dublin, Original Writing for National Association of Principals and Deputies.).

opposite of divergers. They are not very emotional and tend to prefer things to people. They do best in tests that require single solutions. They do best in problem solving in the selection of a solution, and the evaluation of the consequences of that solution. (4) Accommodators are the opposite of assimilators (fourth quadrant). They like doing things and want to devise and implement experiments. They take more risk than those with other learning styles, and excel in those situations where they must adapt themselves to specific immediate circumstances. While being at ease with other people, they are relatively impatient.

In McCarthy's 4MAT model which is an adaptation of the Kolb scheme, Type 1 respondents are divergers who ask "Why Questions." Type 2 are assimilators who ask "What questions." Type 3 are convergers who ask "How questions," and Type 4 are accommodators who ask "What if questions." It follows that students need to learn to ask all four types of question.

Engineering educators have used both the Kolb [25], and 4 MAT approaches to design and implement lessons [26, 27, 28] (see Exhibit 12.5).

12.6   FELDER-SOLOMON INDEX OF LEARNING STYLES

No other engineering educator has done more to promote the significance of learning styles for engineering educators than Rich Felder. In 1988 Felder and Silverman identified 32 different learning styles, and made recommendations about teaching techniques that would address these styles [29]. Like Grasha [30] and Heywood [31] they drew attention to the possible disparities that could exist between student learning styles and engineering teaching styles. They listed five questions, the answers to which would define a student's learning style, and five questions that would define a teacher's learning style (Exhibit 12.6). This is somewhat different from asking teachers to respond to the Kolb Learning Styles Inventory. Their questions were greatly influenced by the Kolb, MBTI (see below) [32] and Witkin learning style theories [33].

The Felder and Solomon Index was developed from this research [34] and continues to be used [35]. Individuals are categorized on a 12 point scale against four dimensions. These are:

Visual/Verbal. Contrasts those who receive information visually with those who prefer verbal explanation.

Sequential/Global. Contrasts those who like a step-by-step presentation of knowledge with those who like knowledge to be "presented in a broad potentially complex manner that allows them to fill in blanks through ah-ha moments."

Active/reflective. Contrasts those who receive knowledge through hands-on activities while internal reflection drives the reflective learner. In the Kolb model, the hands on learner is one who learns through concrete experiences.

Sensing/Intuitive. Contrasts those who like factual knowledge and experimentation with those who like theories and principles. This comes from the Jungian model developed by Myers-Briggs.

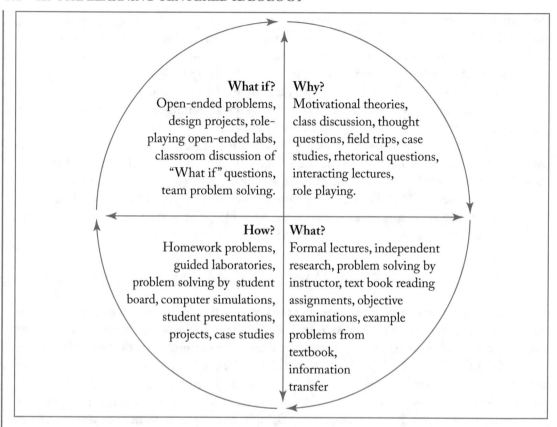

Exhibit 12.5: Todds, adaptation of Svinicki and Dixon's of the Kolb model to teaching showing the 4-Mat questions (Todd, R. H. (1991). (Teaching an introductory course in manufacturing processes. *Engineering Education*, 81(5), pp. 484–485. See also Svinicki, M. D. and N. M. Dixon (1987). The Kolb model modified for classroom activities. *College Teaching*, 35(4), pp. 141–146.).

12.7   TEMPERAMENT AND LEARNING STYLES

There has been a lot of work on the relationship of personality and performance which is by no means conclusive. In engineering there has been particular interest on Jung's extravert and introvert typology. Briggs and Myers designed a questionnaire to elicit the preferences that respondents have for all the psychological types described by Jung [36]. The Myers Briggs Type Indicator (MBTI) was promoted in engineering education by McCauley [37], and has been widely used [38]. McCauley showed the importance of feelings and argued that people skills were undervalued by engineering educators. Feelings have been shown to be important among Chinese engineering students [39].

Questions, the answers to which help define student learning and teacher styles **Questions to students.**	**Questions to teachers.**
1. What type of information does the student preferentially perceive; sensory (external)-sights, sounds, physical-possibilities, insights, hunches?	1. What type of information is emphasized by the instructor: concrete-factual, or abstract-conceptual, sensations, or intuitive (internal) theoretical?
2. Through which sensory channel is external information most effectively perceived: visual-pictures, diagrams, graphs, demonstrations, or auditory-words, sounds?	2. What mode of presentation is stressed: visual-pictures, diagrams, films, demonstrations, or verbal lectures, readings, discussions?
3. With which organization of information is the student most comfortable: inductive-facts and observations are given, underlying principles are inferred, or deductive- principles are given, consequences and applications are deduced?	3. How is the presentation organized: inductively—phenomena leading to principles, or deductively—principles leading to phenomena.
4. How does the student prefer to process information; actively or discussion, or reflectively through introspection?	4. What mode of student participation is facilitated by the presentation: active-students talk, move, reflect, or passive students watch and listen?
5. How does the student progress toward understanding: sequentially in continual steps, or globally in large jumps (holistically)?	5. What type of perspective is provided on the information presented: sequential- step by step progression (the trees), or global context and relevance (the forest)?

Exhibit 12.6: Questions to assist in the identification of learning styles. Felder, R. M. and L. K. Silverman (1988). Learning and teaching styles in engineering education. *Engineering Education*, 78, pp. 674–681 (Reproduced by kind permission of R. M. Felder).

Yokomoto on the grounds that "sensing" types learn best when the material is based on experience and proceeds step-by-step with examples and hands on activities, taught his "sensing" students to master specific examples and then look for connections and patterns. In contrast he taught "intuitive" students to master the mechanics in order to solve problems quickly rather than being brought to a halt when they had grasped the problems. Other research by Yokomoto and his colleagues affirmed his view that different types responded better to different modes of teaching [40].

At the University of Salford the MBTI was used as a diagnostic instrument in a bridging course (i.e., from technician courses to degree level work). Its intention was to help non-traditional university students gain an understanding of their natural learning styles and for them to develop effective learning strategies. Compared with traditional entry students, it was found that the non-traditional students in this innovative course had higher levels of extraversion and feeling. Another test showed that they tended to converge on the problem rather than explore other possibilities [41].

The MBTI became a test that was favored by personnel selectors in industry for which reason Kline (an expert in psychometrics), considered it in his analyses of tests in spite of its perceived flaws [42]. Whatever the problems with this instrument it can lead a teacher to reflect on the methods of instruction used.

Recently however the Five Factor Personality Inventory, known as the Big 5, has been thought to be a better instrument than the MBTI. It has shown a group of Dutch engineers (with a mean age of 48.4 years) to be somewhat more extraverted than the population as a whole, yet, paradoxically they were more autonomous and less friendly than ordinary people which might be problematic in interpersonal relations. They would need to learn to be more "agreeable" (agreeable/quarrelsome being one of the five factors [43]. The authors pointed out that their findings had some similarity with a study of Chinese engineers who were found to be more emotionally stable and conscientious than a comparison group [44]. The Big Five's domains are extraversion, agreeableness, conscientiousness, emotional stability and openness [45].

It is reasonable to conclude that there is substantial evidence to show that studies of learning styles and the temperaments of students can provide educators with insights into student learning, as well as into their methods of instruction [46]. Do you believe students cry "Please understand me," as the title of the book by Kiersey and Bates asks?

NOTES AND REFERENCES

[1] Schiro, M. (2013). *Curriculum Theory. Conflicting Visions and Enduring Concerns*, 2nd ed., Los Angeles. Sage. 153, 155

[2] Plowden, B. (1967). (Chairman of a Committee). *Children and their Primary Schools*. A Report of the Central Advisory Council for Education. London, HMSO. 153

[3] Cowan, J. (2006). *On Becoming an Innovative University Teacher: Reflection in Action*. Maidenhead, SRHE/Open University Press. 153, 164, 167

[4] Borglund, D. (2007). A case study of peer learning in higher aeronautical education. *European Journal of Engineering Education*, 32(1), pp. 35–42. 154

[5] *loc.cit.* ref. [3]. 154

[6] Davies, J. W. and U. Rutherford (2012). Learning from fellow engineering students who have current professional experience. *European Journal of Engineering Education*, 37(4), pp. 354–365. 154

[7] Wigal, C. M. (2007). The use of peer evaluation to measure student performance and critical thinking ability. *ASEE/IEEE Proceedings Frontiers in Education Conference*, S3B, pp. 7–12. 154

[8] Ellis, G., Moriarty, M., and G. Felder (2008). A learner-centered approach to preparing at-risk students. *Proceedings of the Annual Conference of the American Society for Engineering Education*. Paper 1983. 154

[9] Heywood, J. (2000). *Assessment in Higher Education. Student Learning, Teaching, Programmes and Institutions*. London. Jessica Kingsley, pp. 374–378. 154, 172

[10] Brown, P. C., Roediger III, H. L., and M. A. McDaniel (2014). *Make it Stick. The Science of Successful Learning*. Cambridge, MA, Belknap Press, pp. 125–126, and pp. 230–231. 154

[11] White, A., Livesay, G., and K. C. Dee (2017). *Proceedings of the Annual Conference of the American Society for Engineering Education*. Paper 1259. 154

[12] Wenger, F. (1999). *Communities of Practice: Learning, Meaning, and Identity*. Cambridge, Cambridge University Press. 154

[13] Hawkins, J. D., Catalano, R. E. and associates. (1992). *Communities that Care. Action for Drug Abuse Prevention*. San Fransisco, Jossey Bass. 154, 170

[14] Heywood, J. (1972). Short courses in the development of originality in S. A. Gregory (Ed.), *Creativity and Innovation in Engineering*. London, Butterworths. 155

[15] Peters, A.-K. and A. Pears (2012). Student's experiences and attitudes towards learning computer science. *ASEE/IEEE Proceedings Frontiers in Education Conference*, 8. 155

[16] Peters, A. K. Berglund, A., Eckerdal, A., and A. Pears (2014). First year computer science and IT students' experience of participation in the discipline. *International Conference on Teaching and Learning in Computing and Engineering*, IEEE Publications DOI 10.1109/LaTICE.2014.9. 155

[17] Brown, C., Roediger III, H. L., and M. A. Daniel (2014). *Make it Stick. The Science of Successful Learning*. Cambridge, MA, Belknap Press. 155

"We are poor judges of when we are learning well and when we're not. When the going is harder and slower and it doesn't feel productive, we are drawn to strategies that feel more fruitful, unaware that the gains from these strategies are often temporary. Rereading text and massed practice of a skill or new knowledge are by far the preferred study strategies of learners of all stripes, but they're also among the least productive" (p. 3). These writers are referring to mental skills of a specific kind. Physical skills such as planing a piece of wood flat require practice, as does the combination of physical and mental skill used in playing a piano.

[18] Felder, R. M. and R. Brent (2005). Understanding student differences. *Journal of Engineering Education*, 94(1), pp. 57–72. 156

[19] Heywood, J. (2005). *Engineering Education. Research and Development in Curriculum and Instruction.* Hoboken, NJ, IEEE/Wiley. Chapter 5. 156

[20] Freeman, J., McComiskey, J. G., and D. Buttle (1968). Research into convergent and divergent thinking. *International Journal of Electrical Engineering Education*, 6, pp. 99–108. 156

[21] Whitfield, P. R. (1975). *Creativity in Industry*, Harmondsworth. Penguin. 156

[22] Kolb, D. A. (1984). *Experiential Learning. Experience as the Source of Learning and Development.* Englewood Cliffs, NJ, Prentice Hall. 157

[23] Heywood, J. (2008). *Instructional and Curriculum Leadership. Toward Inquiry Oriented Schools.* Dublin, Original Writing for National Association of Principals and Deputies. It is recoded on pages 223 and 224 that: 158

"In order to answer the question, 'can learning styles help us to better understand our students?' we asked our post-graduate student teachers to familiarise themselves with Kolb' theory through lectures and papers by Grasha, and Svinicki and Dixon (i). They were asked to devise a lesson plan that would take the pupils through each quadrant of the cycle. In order for them to devise a test it was suggested that rather than teach a concept in different ways in the four phases of the learning cycle they should advance the lesson through the stages of the cycle. In that way they would be able to test each quadrant. They would also be able to get some idea if there was a correlation between style and performance in the different quadrants (i.e., do the divergers do best in the in the divergent part of the test, the assimilators in the assimilator part of the test and so forth). A week after the lesson was completed the test was to be set. These graduate student teachers were then given additional literature to discuss the question—should learning styles be matched to teaching styles (their response to take into account their own learning styles as assessed in another part of their training programme?). To achieve these goals the students had to administer the Learning Styles Inventory to their students. In addition to the reports many of the student voluntarily completed a questionnaire designed to elicit their opinions about certain dimensions of the exercise, and simplify the analysis of the reports" (ii).

In the text data from two groups of student teachers separated by 5 years was presented, and it was reported that the comments and responses covered the same areas in the two years. Moreover, the examples in the text taken from one of the intervening years support those conclusions. It is of some significance that the first group used the Kolb inventory whereas the second group assigned their own descriptions to their pupils on the basis of the literature which at that time also included the Honey and Mumford Inventory (iii). (We had had to give up the Kolb Inventory because of cost in favour of the much cheaper Honey and Mumford model).

It is reported (pages 224–227) that "most of the students tried to design a lesson in which they could cover the whole of the cycle but there were a few who took two lessons, and one or two who took three or four, each lesson being devoted to one style. One of the latter is shown in Exhibit 12.7 […]. It should be noted that half of the student teachers found it difficult to cover the cycle in a single lesson. No reason was found to suggest that the cycle should not take more than one lesson."

Concerning the question of point of entry into the cycle, "one or two of the students started at different points such as beginning with a theory. Cowan's supports an approach that allows entry at other appropriate points—that is, appropriate to the objectives to be achieved" [3].

Paula Carroll who designed the lesson plan shown in Exhibit 12.7 reported an overall improvement of 8% between pre and post-test scores. In a graded public examination system this is a matter of a difference of a grade. While the material was not totally new to the students, Carroll was of the opinion that this was no bad thing because repetition does not mean assimilation, and the lesson may have helped assimilation. However, she had set a test at the end of each lesson and had expected there to be an improvement in scores as between the first and the last tests. But that had not happened. She put this down to the fact that test 1 was easier than the other tests. She felt that it was notoriously difficult for a novice to design 4 tests of equal validity (p. 224). I think many experienced teachers would have the same difficulty.

Carroll like many other of the graduate student teachers raised the question of the validity of the Kolb Inventory especially when used with younger children (12–13 years of age). Many of these student teachers had to explain some of the items/word in the inventory. "One student teacher who questioned the validity of the inventory administered it twice to her students with a space between. One third of the twenty one tested changed their styles. Inspection of the results suggests that there was a general movement toward the centre of the axes" which was not accounted for her in her comments. But when she came to evaluate her work, she concluded that although there was no evidence to support the view that students learn best in the phase which corresponds to their own learning style, nevertheless teaching a lesson that passes through the Kolb cycle improves learning. This was also the view that Carroll took. She made the point that while students might enter the school as concrete thinkers, they might leave secondary education as convergers. This suggests that the type of instruction might influence the styles of some of the students a point that has been made about convergent and divergent thinking (see above). Carroll also found that while there was no relation between learning style and the phase in which best score was achieved there was a relation between learning style and preferred phase of lesson.

These case studies illustrate the use of inventories as "operators" regardless of what they actually measure. They separate pupils into groups against which the student teacher can reflect on their own teaching and their understanding of their pupils.

(i) Svinicki, M. D. and N. M. Dixon (1987). The Kolb model modified for classroom activities. *College Teaching*, 35(4), pp. 141–146.

(ii) FitzGibbon, A., Heywood, J., and L. A. Cameron (1991). *Matching Teaching Styles to Learning Styles*. Monographs of the Department of Teacher Education. No 1/91. University of Dublin.

(iii) Honey, P. and A. Mumford (1992). *The Manual of Learning Styles*. Maidenhead. Peter Honey.

They call their styles—Activists, reflectors, theorists and pragmatists.

[24] FitzGibbon, A. (1987). Kolb's experiential learning model as a model for supervision of classroom teaching for student teachers. *European Journal of Teacher Education*, 10(2), pp. 163–178. 158

[25] (i) Sharp, J. E., Harb, J. N., and R. E. Terry (1997). Combining Kolb learning styles and writing to learn in engineering classes. *Engineering Education*, 82(2), pp. 93–101. 161
(ii) Sharp, J. E. (1998). Learning styles and technical communication. Improving communication and teamwork skills. *ASEE/IEEE Proceedings Frontiers in Education Conference*, 1, pp. 513–517.

[26] McCarthy, B. (1986). *The 4MAT System. Teaching to Learning Styles with Right and Left Mode Techniques*, Barrington, IL, Excel Inc. 161

[27] Stice, J. E. (1987). Using Kolb's cycle to improve student learning. *Engineering Education*, 77(5), pp. 291–296. 161

[28] Todd, R. H. (1991). Teaching an introductory course in manufacturing processes. *Engineering Education*, 81(5), pp. 484–485. 161

[29] Felder, R. M. and L. K. Silverman (1988). Learning and teaching styles in engineering education. *Engineering Education*, 78, pp. 674–681. 161

Lesson Phases	Content	Learning Strategies
Introduction (Day 1. 1 class period) (1) Students are reminded of Learning Styles Inventory (rationale has been explained) and told they are to have a lesson divided into 4 parts based on LS with a short test after each part. (2) Students divided into 5 groups and told each group will interview the 5 German exchange students present in the school. (3) Groups prepare in German batteries of questions to ask.	(3) Questions prepared are based on topics in the course (e.g., age, hobbies, school, etc. Students are focussed on 1st and 2nd person singular verb forms.	Expository (3) Brainstorming. Some guidance from teacher mainly grammar in question presentation.
Phase 1. Concrete experience (Day 2. 1 class period). (1) The 5 German students are interviewed, one at a time, by each group- such that the group carries out 5 interviews, one with each German student. (2) Students write down information about each German student. (3) Students then given a short verb test.	(1) Questions prepared above in German are used. When exhausted the students switch to English to ask anything of interest to them.	(1) Group Work (Normal interference from teacher) (3) Individual written work.
Phase 2. Reflective observation (Day 3 ¼ of double class period) (1) Discussion/reflection on what students learned in the interviews (in English)	(1) Some leading questions from the teacher in order to direct discussion: *what did you learn about the German students/ anything unusual? *What did you find out about Germany? *Did meeting the people make you interested in the place? *What verbs are you using? In what way did you use them? Focus on verb form+ pronoun for 2nd person singular. *What verbs did the Germans use in answering? The same ones? How did they use them? (focus on verb form+ pronoun for 1st person singular)	(1) Brainstorming. Large group discussion. Guided discovery when focussing on verbs.

Exhibit 12.7: **A Student teachers lesson plan for teaching German verbs using the Kolb Cycle.** (*Continues.*)

(2) Short verb test given. Same format as for phase 1.	(2) See test after phase 2.	(2) Individual work.
Phase 3. Abstract Conceptualization (day 3. 1/3rd Class). (1) Students give written script of an interview with 1st, 2nd, 3rd person singular verb forms on it. (2) They study it and try to work out rules governing use of verb. (3) Short verb test given. Same format as in other phases.	(1) See sheet for phase 3 in attached samples of student work. On it are instructions which guide students in their attempt to work out rules.	(1) Guided discovery: individual or pair work (Students choose how they want to work).
Phase 4. Active experimentation (Day 3. 1/3rd of the class). (1) Students write up report on one of the Germans they interviewed. (2) Short verb test given. Same format as in other phases.	(1) Here 3rd person singular is used	(1) Individual written work.
Conclusion. (Day 3. Final 10 minutes of double class). (1) Students asked to reflect on lesson. (2) Give questionnaires on how they felt about the lesson.		(1) Individual written work. (2) Individual written work.

Exhibit 12.7: (*Continued.*) A Student teachers lesson plan for teaching German verbs using the Kolb Cycle.

[30] Grasha, A. (1984). The journey from Greenwich observatory (1796) to the college classroom. *Improving College and University Teaching*, 32(1), pp. 46–53. 161

[31] *loc. cit.* ref. [13]. 161

[32] For an introduction to the Myers-Briggs Indicator see Kiersey, D. and M. Bates (1984). *Please Understand Me. An Instrument in Temperament Styles*, Oxford, Oxford Psychologists Press. 161

[33] Witkin, H. A. and D. R. Goodenough (1981). *Cognitive Styles*. New York. International Universities Press. 161

Individual dispositions toward the perception of their environment lie on a continuum, the polar ends of which Witkin called field dependent and field independent. Those who are field-dependent look at the world in a global way, while those who are field independent see it analytically. The reaction of the field-dependent person to people, places and events is undifferentiated and complex. In contrast the events (objects) in the environment are not associated with the background of that environment by a person who is field independent.

[34] Soloman, B. S. (1992). *Inventory of Learning Styles*. North Carolina University. 161

[35] Dee, K. C. et al. (2017). Effects of supplemental learning opportunities designed to engage different learning styles. *Proceedings of the Annual Conference of the American Society for Engineering Education*. Paper 1196. 161

McNally, H. (2017). Does the index of learning styles predict laboratory partner success in electronics courses? *Proceedings of the Annual Conference of the American Society for Engineering Education*. Paper 1238.

[36] Myers, I. (1985). *Manual: A Guide to the Development and Use of the Myers-Briggs Type Indicator*. Palo Alto. Consulting psychologists Press. 162

The indicator is based on Jung's theory of personality. According to Jung there are four ways of orienting experience and perceiving the world. *Sensation* results from our sensing the world through our senses to see what exists. *Feeling* is the activity of valuing and judging the world and tells us whether it is agreeable. *Intuition* is perception of the world via the unconscious, and our *thinking* gives meaning and understanding to the world. *Sensing* and *intuition* involve our immediate experiences and are, almost contrary to everyday usage of *feeling* because *feeling* and *thinking* were defined by Jung as rational functions since they require acts of judgement. At the same time the functions are grouped in pairs (i.e., *thinking/feeling* and *sensing/intuition*) and one function is dominant in each pair. Thus, a person who is dominant in *thinking* may have submerged the *feeling* function. One who is dominant in *sensing* may have submerged the *intuition* function. Those functions that are underdeveloped have the power to influence life, and it is from them that strange moods and symptoms emerged. The actualized self requires a synthesis of these four functions.

The Myers-Briggs Type indicator pairs the perception and thinking functions for the purpose of assessing personality and produces four types as follows:

Sensing + Thinking	ST Likely to have career preferences in applied science, business, etc.
Sensing + Feeling	SF Likely to have career preferences for patient care, community service, etc.
Intuition + Feeling	NF Likely to have career preferences for behavioral science, literature and art, etc.
Intuition + Thinking	NT likely to have career preferences for physical science, research, management, etc.

See Engler, B. (1979). *Personality Theories. An Introduction*. Boston, MA, Houghton Mifflin for a study of Jung.

For an introduction to the MBTI see Kiersey, D. and M. Bates (1984). *Please Understand Me. An Essay in Temperament Styles*. Oxford. Oxford Psychologists Press.

[37] McCauley, M. H. (1990). The MBTI and individual pathways to engineering design. *Engineering Education*. 80(5), pp. 535–542. 162

[38] See chapter 9 of Heywood, *loc. cit.* Ref. [9] for a general discussion. 162

Roger Parsons, J. et al. (2017). Comparison of traditional and integrated first year curricula-graduation success and MBTI distribution. *Proceedings of the Annual Conference of the American Society for Engineering Education*. Paper 1089. (Contains a useful primer to the MBTI.)

[39] Zhang, D., Yao, D., Cuthbert, J. N. and S. Ketteridge (2014). A suggested strategy for teamwork teaching in undergraduate engineering programs particularly in China. *ASEE/IEEE Proceedings Frontiers in Education Conference*, pp. 537–544. 162

[40] Yokomoto, C. F., Buchanan, W. W., and R. Ware (1993). Assessing student attitudes toward design and innovation. *ASEE/IEEE Proceedings Frontiers in Education Conference*, pp. 382–385. 163

[41] Culver, R. S., Cox, P., Sharo, J., and A. FitzGibbon (1994). Student learning profiles in two innovative honours degree engineering programmes. *International Journal of Technology and Design Education*, 4(3), pp. 257–288. 164

[42] Kline, P. (2000). *The Handbook of Psychological Testing*, 2nd ed., London. Routledge. 164

He was unable to report that evaluations did bear out the typological claims of the MBTI.

[43] (i) Hendriks, A. A. J., Hofstee, W. K. B., and B. de Raad (1999). The five factor personality inventory. *Personality and Individual Differences*, 27, pp. 307–325. 164
(ii) van der Molen, H. T., Schmidt, H. G., and G. Kruisman (2007). Personality characteristics of engineers. *European Journal of Engineering Education*, 32(5), pp. 495–501.

[44] Dai, X. (2003). A study of the occupational characters of different occupational groups. *Journal of Clinical Psychology*, 10, pp. 252–255. 164

[45] Rhee, J., Parent, D., and C. Oyamo (2012). Influence of personality on senior project combining innovation and entrepreneurship. *International Journal of Engineering Education*, 28(2), pp. 302–309. 164

[46] Rutsohn, J. (1978). Understanding personality types. Does it matter? *Improving College and University Teaching*, 26(4), pp. 249–254. 164

JOURNEY 13

Intelligence

13.1 IQ AND ITS IMPACT

While it may be incumbent on a teacher to have a working view of "intelligence" it is inevitable that most of us will have a theory of intelligence. At one end of the spectrum we may believe that it is innate and genetically conditioned, while at the other end of the spectrum, we may believe that it is a pattern of nurture that is handed down. Not only do our views have a bearing on whom we select for engineering education and subsequently jobs, but how we teach. Linked to our views about intelligence are views about competency, who is competent and how to train for competence.

Nothing is guaranteed to stir the populace more than the findings of intelligence tests reported as a person's IQ (Intelligence Quotient), particularly if they appear to relate to us and our life chances. In the United States a book called "*The Bell Curve*" published by Hernstein and Murray in 1994 caused a furore [1]. Twenty two years later, the result of the American Presidential election in 2016 caused some heads to wonder if its prediction that a cognitive élite would be created, who would work in jobs that keep them away from run-of-the-mill workers, had come true. Similarly, "Does Michael Young's 1958 satire on the English tripartite system of education, '*The Rise of the Meritocracy*,' widely discussed in the US and the UK at the time, now have an element of truth about it?" [2, 3, 4, 5].

In England a tripartite system of education that selected children at age 11, by means of examinations commonly regarded as intelligence tests [6], into one of three types of school, was more or less replaced by non-selective comprehensive schools in the 1970's. But, selection for the 160 high status grammar schools that remained continued. A current proposal (April 2017) from the conservative government to increase the number of grammar school places, in order, so it believes, to increase social mobility, is the cause of a great furore.

This debate in the UK has its origins in a fundamental principle stated in the 1943 Norwood Committee report to the effect, "that the evolution of education has in fact thrown up certain groups, each which must be treated in a way appropriate to itself" [7]. Thus, the English education system had recognized a group of students "interested in learning for its own sake. Its technical education showed that it was important to recognize the needs of the pupil whose interests and abilities lie markedly in the field of applied science and applied art." Finally, there was a group "who deal more easily with concrete things than with ideas." To accommodate these groups, three types of schooling would be required. Grammar schools, technical schools and secondary modern schools. Students would be selected to these schools by means of tests.

Those with the highest scores would be admitted to the grammar schools as a function of the limited number of places available. It was expected that those students would remain in school until they were eighteen, and that many of them would go to university.

There were never enough technical schools to provide for a real tripartite system, so those who did not get into the grammar schools, the majority, went to secondary modern schools which provided an education up to 15+, the age at which compulsory education ceased, for the majority of pupils. The writer of the Wikipedia entry is blunt: The secondary moderns "would train pupils in practical skills aimed at equipping them for less skilled jobs and home management" [8].

This structure had an effect on the image of engineering since students wanting to pursue engineering were held to belong among those suited to technical education. Engineering was clearly thought to be a craft [9]. Thus, the academic-vocational divide persisted.

Supporting the idea of selection by tests was the strong belief in the value and accuracy of psychometric testing.

13.2 PSYCHOMETRIC TESTING

In 1904 a British psychologist Charles Spearman (after whom a coefficient of correlation is named) found that children's scores on tests in different academic subjects were positively correlated. From this data he deduced that each child possessed a general mental ability which he called "g".

Subsequently, there was a search for the mental abilities that made up "g" in both the U.S. and the UK. In 1938 in the United States Louis Leon Thurstone suggested that there were seven such abilities [10]. In the UK Philip Vernon using factor analysis proposed a hierarchical (tree) model at the apex of which was "g" [11]. It was supported by two major group factors (*v.ed*) verbal/educational, and (*k-m*) practical-spatial-mechanical abilities which it might be supposed are important for engineering [12]. Vernon's group factors were supported by numerous smaller specific skills (Exhibit 13.1).

In the United States analysis of the Wechsler Adult Intelligence Test (WAIS-III) also showed a hierarchy among the 13 tests [13]. Later in 1993 John Carroll reported on a meta-study of some 400 data sets which suggested that there were eight types of mental ability similar to the group factors [14].

Deary writes, "Something like John Carroll's three stratum model almost always appears from a collection of mental tests. A general factor emerges that accounts for about half of the individual differences among the scores for a group of people, and there are group factors that are narrower abilities, and then very specific factors below that." He goes on to say, "We can nowadays describe the structure of mental test performance quite reliably, but this is not proven to represent a model of the organization and compartments of the human brain" [15].

Tests of general mental ability are found to be relatively good predictors of job performance. Deary writes "No it will not predict all that strongly how well people do a job. Yes, you

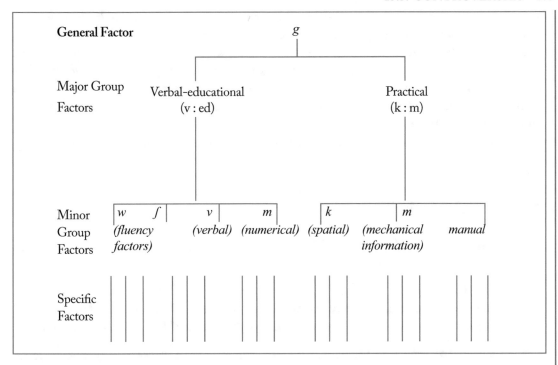

Exhibit 13.1: Hierchical structure of human abilities.

will hire people who are hopeless and with whom you can't get on. But on the whole you'd be better off including a general mental ability test in your portfolio of selection methods" [14], which is a reminder that multiple methods of assessment are to be preferred to a unitary instrument.

13.3 CONTROVERSIES

While Intelligence tests have proved to be relatively good predictors of academic potential and work performance they have nevertheless generated controversies, of which two are relevant to this discussion.

The first controversy, the idea that intelligence is a single entity, (the psychometric view), is challenged by those who think that intelligence is multi-faceted. Probably the best known promoter of this view is Howard Gardner whose 1983 book "*Frames of Mind. The Theory of Multiple Intelligences*" became popular reading. It was very attractive to primary (elementary) school teachers because the curriculum could be designed to develop the seven "contents." This issue will be discussed in more detail in Journey 15.

The second controversy, relates directly to the learner centered and social reconstruction ideologies for at the extreme end of the spectrum is the view that intelligence is inherited. It is part of our nature, and this thinking carries with it the fear that its logic might lead to eugenics. Lewis M. Terman who developed the Binet test in America (the Stanford-Binet Intelligence Scales) said in 1917 that "If we would preserve our State for a class of people worthy to possess it we must prevent, as far as possible the propagation of mental degenerates." A view that is not shared by very many psychologists today. The learner centered and social reconstruction ideologies take the opposite view. But the nature/nurture debate also has a bearing on how we think about competency.

The next chapter will discuss the nature-nurture controversy and its bearing on competency.

NOTES AND REFERENCES

[1] Hernstein, R. J. and C. Murray (1994). *The Bell Curve. Intelligence and Class Structure in American Life*. New York. Free Press. 173

Notwithstanding the row that it provoked, Professor Ian Deary considers that if gave "some of the clearest accounts of statistical analyses" that he had ever read. He considered that the book as a whole was "extremely easy to understand" [2].

There are two aspects to the book. First, it is argued that an intellectual élite is emerging in the United States which is accompanied by a class system based on IQ. They argue that those with high levels of IQ will be on average more productive than those with low level of IQ in the same occupation. They further argued that an IQ score is a better predictor of productivity than a job interview. The economic pressure will be on employers to hire those with the highest IQ. One effect is that the cognitive élite will work in jobs that keep them away from run-of-the-mill workers.

Hernstein and Murray argued that the proportion of crime, poverty, illegitimacy, welfare dependency and unemployment is highest among those with low IQ's. Since, their fertility rates are also higher this serves to bring down the average IQ of Americans as a whole. Link this to their contention that IQ is genetically based and that affirmative action programmes have not been successful in a year when Republicans were seeking to win Congress with proposals that would reduce these programmes then the coals were laid for a heated controversy.

The Editor of *The New Republic* agreed to publish an article of theirs which summarised parts III and IV of the book. However, the associate editors and staff of the journal refused to countenance its publication because of the controversial nature of its content. Eventually it was agreed that they and others could publish critical comments. In consequence most of one issue was devoted to this controversy. There followed hundreds of critical publications.

The editorial that accompanied this issue is worth reading for its contribution to the debates on press freedom on the one hand and academic freedom on the other hand.

[2] Young, M. (1958). *The Rise of the Meritocracy*. Harmondsworth, Pelican. 173, 176

A satire that predicts that political influence will be a function of the intellectual achievement of the individual. That is, persons with very high IQ will become the political class. Unfortunately while he intended that the term should have negative connotations it was used by others, particularly by the British Prime Minister Tony Blair who eulogised the idea of meritocracy.

In an article in *The Guardian* Young wrote "It is good sense to appoint individual people to jobs on merit. It is the opposite when those who are judged to have merit of a particular kind harden into a new social class without room in it for others" [4].

Wiley's published a substantial discussion on the proposition in 2006 [5].

The results of the referendum on Europe in the UK, and the Presidential election in the U.S. brought into sharp focus the proposition that this social structure may have arrived.

[3] Toby Young, Michael Young's son asked this question in a BBC radio programme. *The Rise of the Meritocracy*. BBC Radio 4 (30 Minutes). April 11, 2017. 173, 181

[4] Young, M. (2001). Down with Meritocracy. *The Guardian*, June 29. Guardian News and Media Ltd. Retrieved April 15, 2017. 173, 177

[5] Dench, G. (Ed.), (2006). *The Rise and Rise of Meritocracy*. Chichester, Wiley-Blackwell. 173, 177

[6] They were not intelligence tests as such. The three tests were mathematical ability, general reasoning, and the ability to write an essay on a general topic. They were commonly referred to as IQ tests, Entrance exams, but more often than not as the 11+. 173

[7] Norwood, Sir Cyril (Chairman of a Committee) (1943). *Curriculum and Examinations in Secondary Schools*. Report of the Committee of the Secondary School Examination Council, London, HMSO, p. 2. 173, 178

This built on the work of the Spens Committee (1938) (i), and the Hadow Committee (1931) which had recommended that the system of education be split into separate stages at age eleven or twelve (ii).

(i) Spens, Sir William (Chairman of a Committee) (1938). *Secondary Education with Special Reference to Grammar Schools and Technical High Schools*. Report of the Consultative Committee to the Board of Education, London. HMSO.

(ii) Hadow, Sir Henry (Chairman of a Committee) (1931). *The Primary School*. Report of the Consultative Committee to the Board of Education, London.

[8] Tripartite System of education in England, Wales and Northern Ireland. Retrieved (April 16, 2017). `http://en.wikipedia.org/w/index.php` 174

[9] *loc. cit.* ref. [7, p. 3]. "the various kinds of technical schools which were not instituted to satisfy the intellectual needs of arbitrarily assumed group of children, but to prepare boys and girls for taking up certain crafts-engineering, agriculture and the like. Nevertheless it is usual to think of the engineer or other craftsman as possessing a particular set of interests or aptitudes by virtue of which he becomes a successful engineer or whatever he may become." 174

[10] Thurstone, L. L. (1938). *Primary Mental Abilities*. Chicago, Chicago University Press. 174

Thurstone (originally a mechanical engineer) designed the Primary Mental Ability Test, the purpose of which was to provide a profile of the child's abilities in the spatial, perceptual, numerical, verbal, memory, verbal fluency and inductive reasoning dimensions. It is classed as an ability test and not an intelligence test although instructions for estimating the score of general intelligence are given.

[11] Vernon, P. E. (1950). *The Structure of Human Abilities*. London, Methuen. 174

Vernon defined intelligence thus, "Intelligence A is the basic potentiality of the organism, whether animal or human, to learn and to adapt to its environment…Intelligence A is determined by the genes but is mediated mainly by the complexity and plasticity of the central nervous system…Intelligence B is the level of ability that a person actually shows in behaviour-cleverness, the efficiency and complexity of perceptions, learning, thinking, and problem solving. This is not genetic…Rather, it is the product of the interplay between genetic potentiality and environmental stimulation…I have suggested that we should add a third usage to Hebb's Intelligence A and B, namely Intelligence C which stands for the score or IQ obtained from a particular test" (i).

(i) Vernon, P. E. (1979). *Intelligence, Heredity and Environment*. San Fransisco, W. H. Freeman and cited by `http://www.inteltheory.com` retrieved April 15, 2017.

[12] MacFarlane Smith, I. (1964). *Spatial Ability*. London, University of London Press. 174

The first psychologist to high light the relevance of spatial ability to engineering design seems to have been MacFarlane Smith. In 1964 he related it to the perceived shortage of engineers in the UK to the failure of schools to develop skills in Vernon's (k-m) group. The neglect of spatial ability was the reason why many able students were not interested in science of technology. He also argued that mathematical ability was different to numerical ability and depended on spatial ability. In terms of the cognitive science available at the time it was held that one side of the brain (left) was neglected at the expense of the other (right). The relevance of spatial ability and visualization to engineering design should be self-evident (i).

(i) See for example Chapter 5.4 of Heywood, J. (2005). *Engineering Education. Research and Development in Curriculum and Instruction*. Hoboken, NJ, IEEE Press/Wiley.

Snyder, M. E. and M. Spenker (2014). Assessment of students' changed spatial ability using two different curriculum approaches: Technical drawing compared with innovative product design. *Proceedings Annual Conference of American Society for Engineering Education*. Paper 9841.

Sorby, P., Casey, P., Veurink, and A. Delane (2012). The role of spatial training in improving spatial and calculus performance in engineering students. *Learning and Individual Differences*, 26, pp. 20–29.

[13] Wechsler, D. (1997) *Manual for the Wechsler Adult Intelligence Scale-III*. New York, Psychological Corporation. 174

WAIS-III is described and its hierarchy mapped in Deary ref. [15]. The 13 tests are vocabulary, similarities, information, comprehension, picture completion, block design, picture arrangement, matrix reasoning, arithmetic, digit span, letter-numbering sequencing, digit-symbol coding, and symbol search.

[14] Carroll, J. (1993). *Human Cognitive Abilities. A Survey of Factor Analytic Studies*. Cambridge, Cambridge University Press. 174, 175

The eight 2nd tier factors are Broad visual perception, Broad Auditory perception, Broad retrieval ability, Broad cognitive speediness, Processing speed, General memory learning, Crystalized intelligence and Fluid intelligence.
The idea of Crystalized and Fluid intelligence comes from Raymond Cattell's theory of intelligence (i). He had come to the conclusion that there might be two factors instead of one. He called them "fluid" and "crystallised" intelligence. Kline (ii) reports that these two factors are the largest of the second-order factors, and account for much of the variance in tests of ability. Fluid intelligence is a basic reasoning ability which reflects the flow of information for the brain. Tests for fluid intelligence require the mental manipulation of abstract symbols Crystallised intelligence reflects the culture in which we live. Tests for crystallised intelligence contain vocabulary, general information and reading comprehension.

(i) Cattell, R. B. (1987). *Intelligence. Structure Growth and Action*. Amsterdam, Elsevier.
(ii) Kline, P. (1993). *Handbook of Psychological Testing*. London. Routledge (2nd ed., published in 2000).

[15] Deary, I. J. (2001). *Intelligence. A Very Short Introduction.* Oxford, Oxford University Press. 174, 179

Includes an excellent advisory list of references that cover the debate. Since it was written our understanding of the relationship between intelligence and genetics has considerably advanced.

JOURNEY 14

Two Views of Competency

14.1 NATURE VS. NURTURE: NATURE AND NURTURE

Shortly after Spearman had suggested *"g"* Cyril Burt reported in 1909 that, in the UK, upper class children at private schools did better than those in state elementary schools (lower classes). He became convinced that the children from the lower classes required a different education to those with higher levels of intelligence, and that selection to schools using intelligence tests should be the norm. As we have seen his views had an enormous effect on the organization of schooling in England and Wales. Coupled with this view he believed that intelligence is innate and cannot be altered very much by schooling. Since then there has been much research and debate about the extent to which the environment contributes to the development of intelligence. Clearly, this has implications for the curriculum and instruction as well as the organization of schooling.

The finding that Intelligence correlates with social class caused questions to be asked about the fundamental nature of "intelligence" on which selection to schooling was based [1]. The tests that were used were criticized because a student could be coached to pass them. In England parents not only pay for their children to be coached so that they will get into the nearest of the remaining grammar schools, but they move their homes so as to be in the catchment area of a grammar school, or a good comprehensive school.

A quite serious objection made against testing is that children should not experience "failure" at such an early age. Numerous children, it is argued are socially deprived. The key question has become, "What organizational structure of schooling will provide for all children to have an equal opportunity of success?" For many people in England that answer lies in comprehensive schools. The position we take on the nature-nurture debate has an important bearing on what we mean by "equal opportunity."

There have been equally ferocious debates in the United States about heritability and social deprivation particularly in the inner cities [2].

Clearly, it is in every ones interest to establish the extent to which intelligence is heritable. P. E. Vernon came to the conclusion that 60% of the variance in intellectual ability was attributable to genetic contributions [3]. In the radio programme noted in Journey 13, ref. [3], Robert Plomin suggested that the level was around 50%. He has shown that genetic factors can mediate associations between environmental measures and developmental outcomes. He has argued that we should think about "Nature and Nurture" not "Nature versus Nurture," and that is probably how we should come to think of "inside" and "outside" competencies [4].

14.2   INSIDE AND OUTSIDE COMPETENCIES

Just as engineering educators should have a view about intelligence so they should have a view about competence.

The intention here is not to discuss relationships between competency and intelligence but to draw attention to the fact that there exists among the professions, as for example medicine, a debate about whether competency is something innate that can be affected by instruction, or something that is context dependent. Just as I mostly ignored discussion of the meaning of intelligence in the previous paragraphs, so I intend to ignore discussion of the meaning of competency, about which there is a similar debate. My intention is to draw attention to the key issue that should influence debates about whether or not individuals should be trained to be immediately useful in industry on completion of their education.

Griffin [5] points out that the "inside" view of competency is deeply embedded in the Western psyche. The psychological concepts associated with this model arise from a deeply entrenched cultural model of intelligence [6] which make it difficult for those that hold it to change their attitudes toward teaching and assessment [7]. Griffin writes, "Machine metaphors, common in western conceptions of the mind and thinking, also define what is involved in being a competent person. In many European and American cultural contexts, the person is represented and realized as a separate bounded, autonomous entity, that is, an individual. Individual actions result from the attributes of the person that are activated and, then cause behavior. Accordingly, competence is "located in the individual." Individual actions result from the attributes of the personal that are activated and then, cause behavior. Accordingly, competence is located "in" the individual, "in" the mind, "in" the brain." It follows that competencies can be taught. It is a view that can be traced back to the ancient Greeks, and those who hold it, probably also hold to the scholar academic ideology.

Griffin argues that inside theories of competence produce inside models of assessment with emphasis on performance rather than competence *per se*. The implications for the curriculum and instruction are profound. The debate may be illustrated by reference to issues surrounding the concept of "cultural competence" [8]. Engineers are now expected to be culturally competent, but what does that mean? Does it mean the possession of knowledge that will enable a student to answer survey questions such as "people in the U.S. and India would define an engineering problem similarly?" [9] If so, does it imply that this knowledge adequately prepares a student to work in India, or is that an impossible task? To put it in another way, should industry allow that a person moving to a new culture should be allowed time to become embedded in its engineering culture?

It also follows, that provided you know what competencies industry requires, as for example the ability to communicate, you teach people how to communicate. But that only holds if there is an understanding of what communication entails. Trevelyan argues that engineering educators don't have that understanding because they do not understand the context (environment) in which communication takes place [10].

Trevelyan's work is supported by other studies of engineers at work. For example, in a study of engineers at a Volvo plant in Sweden Sandberg found that how an individual demonstrates competence at work depends on how that individual perceives the task to be accomplished [11]. Competence is possessed at different levels depending on how individuals view the task. Therefore, before training needs can be identified the task has to be done, observed and analysed. Sandberg's finding suggests that training should become more person (learner) centered. But the more important point is that competence develops within the industrial context as a function of the interaction between the task and the individual.

Blandin in another study found that the core competence at work which he called "acting like an engineer" took a long time to develop. It could not be developed without experience in the company [12]. Blandin argued that there were three levels at which competence functions that need to be understood. These are at the level of the individual, at the level of the group in which an individual works, and at the level of the organization in which the individual works.

These findings challenge the notion that a university can prepare a student to be immediately available for work. The student will have to learn at work. It needs to be remembered that learning is continuous, and that organizations can enhance or impede that learning [13, 14]. There do not seem to be many attempts to provide students with learning how to learn courses, or by course tutors, although at least one report shows how such a course can help with the transition from high school to college [15]. That said, note needs to be taken of bridging courses that seek resolve this particular industry skills gap. They need to take into account how people develop and acquire competence. Clearly, project management cannot be learned except as a project manager, but can a university prepare a person for project management? In this respect the work of the Midwest Coalition for Comprehensive Design Education is of considerable interest [16]. At the same time there is evidently need for more studies of the kind undertaken by Blandin and Sandberg.

Whether industry likes it or not, it necessarily has a role in the development of competence, and it is from that position that discussion about the engineering curriculum should begin. That is, the curriculum should be perceived in terms of intellectual and personal development that continues throughout life. And, that places considerable responsibility on industry for the development of the individual which most organizations do not seem to accept.

NOTES AND REFERENCES

[1] Floud, J., Halsey, A., and F. Martin (1953). Educational opportunity and social selection in England. *Transactions of Second World Conference of Sociology*, II, pp. 194–208. 181

[2] For example the debate triggered by the publication of R. J. Hernstein and C. Murray in *The Bell Curve. Intelligence and Class Structure* (1994). New York, Free Press. See issue of *The New Republic*, 1994 devoted to criticisms of this book. See also a controversy caused

by A. R. Jensen on the *Harvard Educational Review*. See vol. 39 issue 3, 1969 for the comments on Jensen's article which is in vol. 39 issue 1. See Journey 13. 181

[3] Vernon, P. E. (1979). *Intelligence: Heredity and Environment*. San Fransisco, W. Freeman. 181

[4] Plomin, R. et al. (1997), *Behavioural Genetics*, 3rd ed., New York, W. H. Freeman. 181

[5] Griffin, C. (2012). A longitudinal study of portfolio assessment to assess competence of undergraduate student nurses. Doctoral dissertation. Dublin, University of Dublin. 182

[6] Epstein, R. M. and E. M. Hundert (2002). Defining and assessing professional competence. *Journal of American Medical Association*, 287(2), pp. 226–235. 182

[7] Bielefeldt, A. (2017). Cultural competency assessment. *Proceedings of the Annual Conference of the American Society for Engineering Education*. Paper 2313. 182

[8] *ibid* 182

[9] Heywood, J. (2016). *The Assessment of Learning in Engineering Education. Practice and Policy*. Hoboken, NJ, IEEE Press/Wiley. 182

[10] Trevelyan, J. (2014). *The Making of the Expert Engineer*. London, CRC Press/Taylor and Francis. 182

[11] Sandberg, J. (2000). Understanding human competence at work. An interpretive approach. *Academy of Management Journal*, 43(30), pp. 9–25. 183

[12] Blandin, B. (2011). The competence of an engineer and how it is built through an apprenticeship programme. A tentative model. *International Journal of Engineering Education*, 28(1), pp. 57–71. 183

[13] Youngman, M. B., Oxtoby, R., Monk, J. D., and J. Heywood (1978). *Analysing Jobs*. Aldershot, Gower Press. 183

[14] Heywood, J. (2016). The idea of a firm as learning organization and its implications for learning-how-to-learn. *Philosophical and Educational Perspectives in Engineering and Technological Literacy*. Handbook 3. Washington DC, American Society for Engineering Education. A Handbook of the TELPHE Division. 183

[15] Rose, A. and K. S. Kinsinger (2017). Incorporating diversity and international awareness into an introduction to engineering technology seminar course. *Proceedings Annual Conference American Society for Engineering Education*. Paper 2907. 183

[16] Jovanovic, V. and M. Tomovic (2017). A Competency gap in the comprehensive design education. *Proceedings of the Annual Conference of the American Society for Engineering Education*. Paper 664. 183

JOURNEY 15

From IQ to Emotional IQ

15.1 INTRODUCTION

In 1983 Howard Gardener proposed a theory of multiple intelligences in which he suggested that instead of one intelligence there were many. He defined intelligence as "the ability to solve problems, or to create products, that are valued within one or more cultural settings." This allowed for the cultural element that had been a problem for the testing movement. They had wanted to develop tests that were culture free. He wrote, "genius (and, *a fortiori*, ordinary performance) is likely to be specific to particular *contents* that human beings have evolved to exhibit several intelligences and not to draw variously on one flexible intelligence" [1]. He listed the seven contents or intelligences shown in Exhibit 15.1.

Since Gardener believes that these intelligences relate to particular parts of the brain it is more than probable that developments in neuro-science will either validate or invalidate his theory. In the meantime many individuals will find it an attractive explanation of intelligence, and if not intelligence "talent," and that raises the question, "Is engineering a content or a talent?"

Two years later Robert Sternberg published "*Beyond IQ; A Triarchic Theory of Intelligence*" in which like Gardener he argued that there was more to intelligence than academic intelligence as measured by the psychometricians [2]. His information processing approach sought to understand the mental processes undergone by the respondent in answering questions, the speed and accuracy with which they are carried out, and the types of mental representations of information these processes act upon.

Intelligence has to take into account the experience that a person has for we learn from experience how to adapt to the particular environment we find ourselves in. Thus, in Sternberg's view, intelligent behavior is a "mental activity directed toward purposive adaptation to, and selection and shaping of real world environments relevant to one's life." Sternberg calls the intelligence we use in real world encounters "practical intelligence" [3]. He also believes that intelligence can be taught [4].

Without delving into the theory any further it seems evident that intelligent behavior as described by Sternberg is related to the effective management of oneself and others, as well as to professional competence. The skills in the first component of this three component theory shown in Exhibit 15.2 (panel a) illustrate this point. It is likely, therefore that there will be good agreement between implicit theories of intelligence held by the public, and implicit theories of the competencies that contribute to professional performance.

Gardener's contents of intelligence
1. Linguistic Intelligence. All the skills involved in writing, reading, talking and listening. Everyone is endowed with potential in this area.
2. Musical intelligence. A talent which emerges early but for the majority ceases to develop after school years begin. The environment can enhance or develop this talent. The intelligence involved in singing, playing, conducting and appreciation.
3. Logical-mathematical intelligence. Is involved in mathematical and scientific thinking and numerical computation. More generally in the solving of logical problems.
4. Spatial intelligence. An intelligence which draws together a number of loosely connected abilities related to forming spatial relationships as for example the contrast between an artistic perspective and navigation. It is involved in engineering and scientific activities and a skill in chess playing. In contrast to logico-mathematical intelligence which "concludes its developmental trajectory with increasing abstraction, spatial intelligence remains tied fundamentally to the concrete world…"
5. Bodily-Kinesthetic Intelligence. Completes the trio of object related intelligences. This intelligence is exercised in control of one's body. Involved in acting, athletics, dancing and making things (e.g., as in metalwork and surgery). Such activities are problem solving activities.
6. Interpersonal intelligence. Involved in understanding of other people and one's self in relation to an other.
7. Intrapersonal intelligence. The skill involved in understanding one's self.

Exhibit 15.1: Gardener's seven contents of intelligence.

This chapter, therefore, begins with an examination of what lay people and specialists perceive intelligence to be. It is contrasted with what managers in a variety of occupations expect from graduates they wish to recruit. The skill lists that emerge are remarkably similar from which I conclude that many of us have mental models of intelligent behavior that extend beyond that which is measured by intelligence tests.

15.2 IMPLICIT THEORIES OF INTELLIGENCE, FORMAL, AND UNINTENDED BUT SUPPORTIVE

En-route to the development of his triarchic theory of intelligence in a formal study, Sternberg and his colleagues interviewed nearly 500 lay people from a range of occupations together with 140 researchers who specialized in intelligence for their views about intelligence [5]. The respondents were asked what they thought were the characteristics of "intelligence," "academic intelligence," "everyday intelligence" and "unintelligence." Two hundred a fifty behaviors were recorded of which 170 were defined as characteristics of intelligence. These were then rated by

(a) **Components of Intelligence** 1. Meta-components	(b) Higher order processes called by Sternberg executive processes. Operation of the meta-components involves (a) Recognition of the problem (b) Recognizing the characteristics of the problem (c) Selecting lower order processes to deal with the problem (d) Selecting a strategy into which to combine these components. (e) Selecting a mental representation upon which the components and strategy can act (f) Allocating ones mental resources (g) Monitoring the problem solving as it occurs. (h) Evaluating the process.
2. Performance components.	e.g., in inductive reasoning these are encoding, inference, mapping, application, comparison, justification, and response.
3. Knowledge acquisition components	The subject learns how to deal with these components. These involve elimination of irrelevant information. The utilization of knowledge acquisition components in vocabulary learning situations is critical to the development of intelligence.
(c) **Experience and Intelligence** 1. Ability to deal with novelty 2. Ability to automatize information processing.	(d) Experience is a powerful influence on performance so any assessment of performance has to take into account the level of experience. Complex tasks can only be carried out because many of the relevant operations have been automatized.
(e) **Context of Intelligence** 1. Adaptation to one's environment. 2. Environmental selection. 3. Environmental shaping.	(f) Adaptation takes place in context but sometimes it may be necessary to change that context (environment) in order to better shape it to ourselves.

Exhibit 15.2: Summary of R. J. Sternberg's Triarchic Theory of intelligence. For a useful explanation see The Theory of successful of human intelligence Chapter 2 in Sternberg, R. J., Kaufman, J. C., and E. L. Grigorenko (2008). *Applied Intelligence*. Cambridge. Cambridge University Press.

a small group of people, and the ratings factor analysed. Three factors accounted for nearly 50% of the variance. These are shown together with associated descriptions in Exhibit 15.3

Practical problem solving ability: reasons logically and well, identifies connections among ideas, sees all aspects of a problem, keeps an open mind, responds to other's ideas, sizes up situations well, gets to the heart of the problem, interprets information accurately, makes good decisions, goes to original sources of basic information, poses problems in an optimal way, is a good source of ideas, perceives implied assumptions and conclusions, listens to all sides of an argument, and deals with problems resourcefully.

Verbal ability: speaks clearly and articulately, is verbally fluent, converses well, is knowledgeable about a particular field, studies hard, reads with high comprehension, reads widely, deals effectively with people, writes without difficulty, sets times aside for reading, displays a good vocabulary, accepts norms, and tries new things.

Social competence: accepts others for what they are, admits mistakes, displays interest in the world at large, is on time for appointments, has social conscience, thinks before speaking and doing, displays curiosity, does not make snap judgments, assesses well the relevance of information to a problem at hand, is sensitive to other people's needs and desires, is frank and honest with self and others, and displays interest in the immediate environment.

Exhibit 15.3: Abilities which contribute to intelligence. Obtained from questions about the nature of intelligence, academic intelligence, and unintelligence put to experts in research on intelligence and lay persons by R. H. Sternberg and his colleagues. Among the findings was the fact that research workers considered motivation to be an important function of intelligence whereas lay persons stressed interpersonal competence in a social context. In R. H. Sternberg (1985) *Beyond IQ. A Triarchic View of Intelligence*. Cambridge University Press.

There were two main differences between the experts and the lay people. The experts thought that motivation was an important component of academic intelligence whereas the lay people placed much greater stress on the sociocultural aspects of human behavior, that is, interpersonal competence in a social context. Clearly, this profile extends the notion of intelligence discussed in the previous chapter. It also fits with the mental notions with which persons such as personnel (HR) managers use to select people for work.

In the UK one unintended but supportive study analysed 10,000 advertisements for graduates to see what skills employers sought [6]. 59% contained explicit reference to personal skills required for performance on the job. Of the remainder a further 15% could be inferred to require such characteristics. Of the 32 significant characteristics that were isolated 20 were considered to be genuine transferable skills (see Exhibit 15.4). These collated into the four generic categories of communication; teamwork; problem solving (creativity); and, management and organizing as shown in Exhibit 15.5.

Being assertive
Chairing, clarifying, closing, collaborating, confronting, consulting, contracting, critical thinking
Data handling, decentering, delegating
Empathsizing
Facilitating
Hypothesizing
Information gathering
Integrating, interpreting, interviewing
Leading, listening
Mentoring
Negotiating, non-verbally communicating
Opening
Presenting
Questioning
Reflecting back, reviewing
Self-disclosure, supervising, synthesizing
Telephoning.

Exhibit 15.4: Personal skills identified by the Sheffield personal Skills Unit.

This study which was undertaken by the University of Sheffield's Personal Transferable Skills Unit contributed to the UK Government's Employment Department's (equivalent U.S. Department of Labor) Enterprise in Higher Education Initiative. Its assessment committee listed the four broad areas of learning that are required to equip students for their working lives shown in Exhibit 15.6 [7]. There are remarkable similarities between these unintended studies (that is of intelligence) and the Sternberg outcomes. In this respect the significance of studies of what people do at work and how they do it should be apparent [8, 9].

In a more recent publication Sternberg, Kaufman and Grigorenko have suggested that the list of skills in Exhibit 15.3 can be used as a behavioral check list [10]. They suggest you can rate yourself on a scale of 1 (low) to 9 (high) and judge the extent to which each of these behaviors characterizes your typical performance. They suggest that higher ratings are associated with better performance. The same can be done with the lists that emerge from the unintended studies.

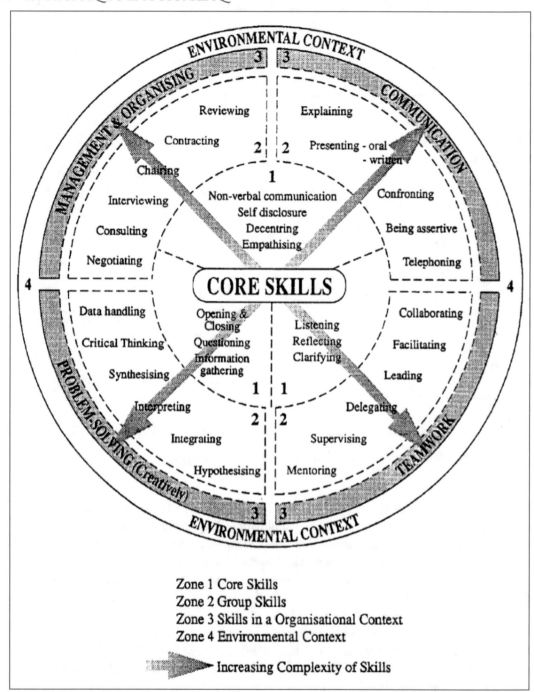

Zone 1 Core Skills
Zone 2 Group Skills
Zone 3 Skills in a Organisational Context
Zone 4 Environmental Context

Increasing Complexity of Skills

Exhibit 15.5: The Personal Skills developmental model described by the Sheffield Personal Skills Unit.

Cognitive knowledge and skills

    Knowledge: Key concepts of enterprise learning (accounting, economics, organizational behavior, inter- and intra-personal behavior).

    Skills: The ability to handle information, evaluate evidence, think critically, think systematically (in terms of systems), solve problems, argue rationally, and think creatively.

Social skills: as for example the ability to communicate, and to work with others in a variety of roles both as leader and team leader.

Managing one's self: as for example, to be able to take initiative, to act independently, to take reasoned risks, to want to achieve, to be willing to change, to be able to adapt, to know one's self and one's values, and to be able to assess one's actions.

Learning to learn: to understand how one learns and solves problems in different contexts and to be able apply the styles learned appropriately to the solution of problems.

Exhibit 15.6: Four broad areas of learning that are important for equipping students for their working lives, as defined by the REAL working group of the UK Employment Department-1991 (cited in Heywood, J. (2005) *Engineering Education. Research and Development in Curriculum and Instruction.* Hoboken, NJ, IEEE/Wiley.

More than that The Sheffield Unit showed how different methods of instruction and learning might achieve the development of the different skills (see Exhibit 15.7). Four active learning situations were compared for the opportunities they afforded for skill development. These were personal tutorials, seminars, project work, and personal profile analysis. The Sheffield unit's staff helped to promote these skills in the learning situations specified in the cells. Workshops and training sessions were also provided. For example, in Medical General Practice tutors identified the skills that were inculcated during small group tutorial work, and they reviewed the extent to which these were made explicit to their students as valuable learning outcomes of their course. In this respect the form of assessment devised by Freeman and Byrne for the assessment of medical students in post-graduate courses in general practice cuts across both cognitive and affective domains [11].

It is clear from the Sheffield analysis that student-led seminars and team projects offer greater scope for skill development than those situations where students work individually with a tutor. This raises the question, "Do engineering educators make the most of the opportunities that are available for personal skill development, and should they?" An attempt will be made to answer this question in the section on emotional intelligence.

Some years earlier Weston and Cranton had related instructional methods to the cognitive, affective and psychomotor domains as shown in Exhibit 15.8 [12]. But this was in 1986. An updated version would include such things as cooperative learning, debates, and assessment

	Communication (C)	Teamwork (T)	Problem-solving (P-S)	Managing and Organizing (M)	Summary of profiles
Personal tutorial	Zone 1 All core skills Zone 2 Explaining, Presenting, Written, Confronting, Being assertive		Zone 1 All core skills	Zone 1 All core skills Zone 2 Reviewing, Contracting, Negotiating	C = 4 T = 0 P-S = 6 M = 2
Seminar tutorial	Zone 1 All Core skills Zone 2 Explaining, Presenting, oral and written, Confronting, Being assertive		Zone 1 All core skills	Zone 1 All core skills	C = 4 T = 0 P-S = 5 M = 0
Seminar. Student-led *Individual*	Zone 1 All core skills Zone 2 Explaining, Presenting, oral and written, Confronting, Being assertive		Zone 1 All core skills Zone 2 Data handling, Critical thinking, Synthesizing, Interpreting, Integrating, Hypothesizing	Zone 1 All core skills Zone 2 Consulting	C = 5 T = 0 P-S = 6 M = 0
Seminar Student-led. *Team*	Zone 1 All core skills Zone 2 Explaining, Presenting, oral and written, Confronting, Being assertive	Zone 1 All core skills Zone 2 Collaborating, facilitating, leading, delegating	Zone 1 All core skills Zone 2 Data handling, Critical thinking, Synthesizing, Interpreting	Zone 1 All core skills Zone 2 Reviewing, Contracting, Chairing, Negotiating	C = 5 T = 4 P-S = 6 M = 4
Project Theoretical e.g., library project, artifacts study. *Individual*	Zone 1 All core skills Zone 2 Explaining, Presenting, oral and written, Telephoning		Zone 1 All core skills Zone 2 Data handling, Critical thinking, Synthesizing, Interpreting, Integrating, Hypothesizing	Zone 1 All core skills Zone 2 Reviewing, Interviewing	C = 4 T = 0 P-S = 6 M = 6

(continued ...)

Exhibit 15.7: Active learning strategies to encourage development in the skill areas shown in Exhibit 15.5. (*Continues.*)

(continued ...)

Project Theoretical, e.g., artifact, survey, experimental practical, fieldwork. *Team*	Zone 1 All core skills Zone 2 Explaining, presenting oral and written, Being assertive, Telephoning	Zone 1 All core skills Zone 2 Collaborating, Facilitating, Leading, Delegating	Zone 1 All core skills Zone 2 Data handling, Critical thinking, Synthesizing, Interpreting, Integrating	Zone 1 All Skills Zone 2 Reviewing, Contracting, Chairing, Negotiating	C = 5 T = 4 P-S = 6 M = 6
Project – "live" work based, clinical placement, company based. *Team*	Zone 1 All core skills Zone 2 Explaining, presenting oral and written, Confronting, Being assertive, Telephoning	Zone 1 All core skills Zone 2 Collaborating, Facilitating, Leading, Delegating, Supervising	Zone 1 All core skills Zone 2 Data handling, Critical thinking, Synthesizing, Interpreting, Integrating, Hypothesizing	Zone 1 All core skills Zone 2 Reviewing, Contracting, Interviewing, Consulting, Negotiating	C = 6 T = 5 P-S = 6 M = 6
Student profile negotiated with tutor	Zone 1 All core skills Zone 2 Explaining, Being assertive		Zone 1 All core skills Zone 2 Critical thinking, Synthesizing, Interpreting, Integrating	Zone 1 All core skills Zone 2 Reviewing	C = 2 T = 0 P-S = 4 M = 1

Exhibit 15.7: (*Continued.*) Active learning strategies to encourage development in the skill areas shown in Exhibit 15.5.

method. Both Sternberg and his colleagues and the Sheffield unit recognized the importance of being able to decode non-verbal cues (behavior).

These studies suggest that something more than the development of academic intelligence is required from the curriculum in Higher Education and Engineering in particular which, while returning us to the position occupied by such specialists in intelligence as Sternberg and Gardener, brings us to the problem of emotional intelligence.

15.3   EMOTIONAL INTELLIGENCE

We have also learned that we need to be able to govern (control) our emotions, and the ability to do this is sometimes called emotional (or social) intelligence. Goleman's (1995) published a book on emotional intelligence was an immediate best seller [13]. In it he asked, "What factors are at play, for example, when people of high IQ flounder and those of modest IQ do surprisingly well?"

Domain and level	
Cognitive domain	
Knowledge	Lecture, CAI, drill and practice
Comprehension	Lecture, modularized instruction, CAI
Application	Discussion, simulation and games, CAI, modularized instruction, field experience, laboratory
Analysis	Discussion independent/group projects, simulations, field experience, role playing laboratory.
Synthesis	Field experience, role playing, laboratory independent/group projects.
Evaluation	Independent/group projects, field experience, laboratory.
Affective domain	
Receiving	Lecture, discussion, modularized instruction, field experience.
Responding	Discussion, simulation, modularized instruction, role playing, field experience.
Value	Discussion, independent/group projects, simulations, role playing, field experience.
Organization	Discussion, independent/group projects, field experience
Characterization by value	Independent projects, field experience.
Psychomotor domain	
Perception	Demonstration (lecture), drill and practice
Set	Demonstration (lecture), drill and practice
Guided response	Peer teaching, games, role playing, field experience, drill and practice
Mechanism	Games, role playing, field experience, drill and practice
Complex overt response	Games, field experience
Adaptation	Independent projects, games, field experience
Organization	Independent projects, games, field experience.

Exhibit 15.8: Matching objective, domain and level of learning to appropriate method and instruction by C. A. Weston and P. A. Cranton (1986) [12].

He went on to argue, "That the difference quite often lies in the abilities called here emotional intelligence which includes self-control, zeal and persistence and the ability to motivate oneself. And these skills […] can be taught to children giving them a better chance to use whatever intellectual potential the genetic lottery may have given them."

The subjects of emotional and social intelligence have been studied during most of the last century, irrespective of whether they are the same construct or different constructs. Taken together, they may be considered as ways of "understanding individual personality and social behavior" [14]. There are non-traditional intelligences such as "practical intelligence" that seem to overlap with them. Hedlund and Sternberg have been concerned to establish if they are overlapping constructs [15]. But, although Sternberg has been involved in the debates about emotional intelligence [16] he does not mention them in either of his books on "*Applied Intelligence*" or "*Wisdom, Intelligence, Creativity and Success*" except for brief references to emotional reasoning.

Inventories have been developed to measure emotional intelligence. Bar-On's has been considered by some experts to be both a personality construct and a mental ability [17]. It is for this reason that much research that has been done has been within the framework of traditional thinking about personality and intelligence, and about the cognitive and affective domains of educational thinking.

When Culver [18], argued that promoting emotional intelligence would be necessary if a successful engineering programme is to be achieved, he cited the list of components that make up emotional intelligence from the "Self-science curriculum used in Nueva School," California [19], and shown in Exhibit 15.9. It might be objected that the Nueva skills are not a distinct social or emotional intelligence but rather a set of personality traits, in which case they are better called personal transferable skills as the Sheffield unit did.

One way of looking at emotional intelligence is to consider it to be the interplay between the cognitive and affective domains in the conduct of living, if you accept that is, that living is problem solving and critical thinking in contexts that always have some emotional context. But, as Hedlund and Sternberg [20] pointed out, the competencies required to solve problems will be a function of the type of problem to be solved, and we might add, therefore individuals need to have a wide range of personal transferable skills. In Sternberg's terms they need to have "practical intelligence."

15.4 PRACTICAL INTELLIGENCE

Sternberg and his colleagues included within the domain of practical intelligence, practical problem solving, pragmatic intelligence and everyday intelligence

"Practical intelligence involves a number of skills as applied to the shaping of and selection of environments" (which Sternberg argued is what intelligent people do). "These skills include among others (1) recognizing problems, (2) defining problems, (3) allocating resources to solving problems, (4) mentally representing problems, (5) formulating strategies for solving problems, and (6) evaluating solutions to problems" [21].

Hedlund and Sternberg considered that what differentiates emotional from social and practical knowledge is "tacit knowledge." That is, the knowledge that is not taught, but acquired as part of everyday living. As Michael Polyani who identified this category of knowledge put it "We know more than we can tell" [22]. The idea is vividly captured in Yorkshire dialect by the term "nouse!" [23]. This knowledge is acquired from managing one's self, managing tasks, and managing others. It is as Trevelyan has shown of major importance in the practice of engineering [24].

A key skill in the development of tacit knowledge is self-reflection yet the Sheffield study found that engineering students do not like to self-reflect. "They were not used to talking in terms of feelings, nor could they see the relevance of such reflection to learning about engineering

Self-awareness	Observing yourself and recognizing your feelings with a view to action or trying to change action in specified circumstances. This can include mode of study, reactions to people, etc.
Personal decision making	Examining one's actions and predicting the consequences. Knowing the basis of the decision, i.e cognition or feeling. This covers the gamut of small and large decisions that relate to everyday actions.
Managing feelings	Requires self-awareness in order to be able to handle anxieties, anger, insults, put-downs, and sadness.
Handling stress	Use of imagery and other methods of evaluation
Empathy	Understanding how people feel and appreciating that in the learning situation students can become stressed and that such stress can be reduced by the mode of instruction (e.g., use of imagery).
Communications	Becoming a good listener and question asker; distinguishing between what someone else does or says and your own reactions about it; sending "I" messages instead of blame.
Self-disclosure	Building trust in relationships and knowing when one can be open.
Insight	This is different from cognitive insight referred to previously. It is about understanding one's emotional life and being able to recognize similar patterns in others so as to better handle relationships.
Self-acceptance	Being able to acknowledge strengths and weaknesses, and being able to adapt where necessary
Personal responsibility	Being able to take responsibility for one's actions. This relates to personal decision making. Learning not to try and pass the buck when the buck really rests with one's self.
Assertiveness	The ability to be able to take a controlled stand. i.e. with neither anger nor meekness. Particularly important in decisions involving moral issues in engineering on which the professional ethic demands that a stand should be made.
Behavior in groups	Knowing when to participate, lead and follow.
Conflict resolution	Using the win/win model to negotiate compromise. This is particularly important in industrial relations and it applies to both partners in managerial conflicts.

Exhibit 15.9: Culver's listing of the Nueva School components that make up emotional intelligence.

problems." Whether or not one should talk about one's feelings has become a matter of debate in the UK since this year Prince Harry revealed the personal difficulties that arose from the death of his mother [25].

"The ability to acquire knowledge, whether it pertains to managing one's self, managing others, or managing tasks can be characterized appropriately as an aspect of intelligence. It requires aspects such as encoding essential information from the environment and recognizing associations between new information and existing knowledge. The decision to call this aspect of intelligence social, emotional, or practical intelligence will depend on one's perspective and one's purpose" [26].

It seems to me that any interaction between individuals involves a degree of emotional contact, sometimes highly charged, that has to be controlled, and that we might better control ourselves and situations if we better understand ourselves and others. That dimension is missing from Sternberg's discussion, but necessarily implied, if not recognized. Similarly, the engineering curriculum pays precious little attention to this dimension even when it teaches communication. It is not sufficient to rely on tacit knowledge because the judgments we make about others can often turn out to be wrong. A teacher is helped if they identify the cognitive-emotional states of their students.

In this respect in 2002 Koort and Reilly of the Media Lab at MIT argued that the ability to be able to identify a learner's cognitive emotional state should enable teachers to provide more efficient and pleasurable learning experiences [27]. They believed that teachers could do this by observing facial expressions, gross body language, and the tone and content of speech. Some teachers make such judgments automatically, but others are insensitive to such situations. Those who are sensitive to them may not know what to do about them in classroom situations. For this reason and for the purpose of training Koort and Reilly offered a four quadrant model that related learning to the emotions. It is shown in Exhibit 15.10. "Similarities" with the Kolb model of learning styles will be apparent.

Koort and Reilly called the vertical axis "the learning axis." Knowledge is constructed in an upward direction and misconceptions are discarded in the downward direction. The model's intention is, on the one hand to show that learning in science, engineering and math is naturally cyclic, and on the other hand, to demonstrate that when students find themselves in the negative half that this is inevitable. Thus, the teacher has to help students to keep orbiting the loop and "to propel themselves, especially after a set-back." The model suggests intervention strategies that the teacher might use in each quadrant.

Taken together Journeys 11, 12, 13 and this chapter demonstrate that not only teaching but policy making in respect of the curriculum benefit if we have a wide ranging understanding of student behavior.

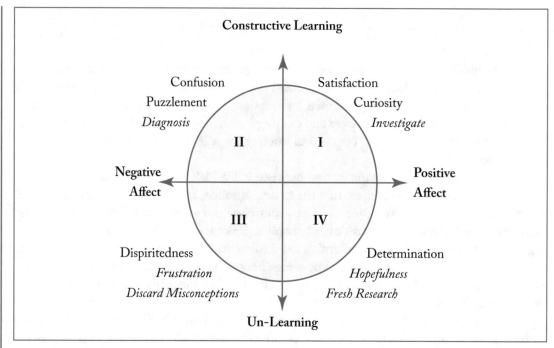

Exhibit 15.10: Relating learning to the emotions from Koort, B. and R. Reilly (2002). A pedagogical method for teaching scientific domain knowledge. *ASEE/IEEE Proceedings Frontiers in Education Conference*, 1, T3A. 13 to 17.

NOTES AND REFERENCES

[1] Gardener, H. (1983). *Frames of Mind. The Theory of Multiple Intelligences.* New York, Basic Books. 185

Gardener stated that he was open to the use of other terms to describe "intelligences" such as intellectual competencies, thought processes, and cognitive skills. This attitude illustrates how his approach differed from the psychometric tradition. Because, he argued about the interaction between heredity and training, we each develop these intelligences—some to a greater extent than others. From the beginning of life these intelligences build upon and interact with each other. At the core of each intelligence is some kind of device for information processing that is unique to that particular intelligence.

Gardener suggests that the presentation of information to the nervous system causes it to carry out specific nervous operations, which, by repeated use and processing, are eventually generated as intelligent knowledge. These "raw" intelligences are developed through their involvement with symbol systems. Children demonstrate their various intelligences through their grasp of these symbol areas. Then with further development, each intelligence and its associated symbol system come to be represented in a notational system. For example, at the symbol system level there is the response to the sound of a musical instrument: at the second order level there is a response to music. Finally, these intelligences are expressed in vocational interests.

Gardener's theory is both open to criticism and development. He would not deny that others are likely to come along and propose other intelligences. For example, is it possible that there is an engineering intelligence? It is questionable whether computational skills should be linked with mathematical ability. But, he does accept that there is a strong association between logico-mathematical ability and spatial intelligence. He has been accused of selecting evidence to suit his case. He is also strongly influenced by the theories of Piaget notwithstanding the criticisms that have been made of Piaget's theories.

Gardener has been anxious to apply his theory in educational settings and some elementary school teachers have designed curricula to meet the requirements of the theory, not always without difficulty (i). As an ideology the examples show that it is learner centred. Doubtless, neuro science will eventually be able to shed light on the validity of the theory. In the meantime the question is whether the idea is of value to engineers. The fact that spatial intelligence is regarded as a separate intelligence raises questions about the attention engineering educators pay to this dimension in the their teaching, as we saw in the last chapter. Industrialists are as much concerned with intrapersonal relations as they are with technical skills. Witness the volume of literature on emotional intelligence (see Section 15.4).

Gardener did not provide conclusive proof of his theory.

The idea of multiple intelligences challenges the view that IQ tests are the best means of sorting people into occupations since they do not measure the range of human potential.

(i) Armstrong, T. (2003). *The Multiple Intelligences of Reading and Writing. Making the World come Alive.* Alexandria, VA, Association for Supervision and Curriculum Development.

Cambell, L. and B. Cambell (1999). *Multiple Intelligences and Student Achievement. Success Stories from Six Schools.* Alexandria, VA, Association for Supervision and Curriculum Development.

[2] Sternberg, R. J. (1985). *Beyond IQ. A Triarchic Theory of Intelligence.* New York. Cambridge University Press. 185

[3] Sternberg, R. J., Kaufman, J. C., and E. L. Grigorenko (2008). *Applied Intelligence.* Cambridge, Cambridge University Press. 185, 200

[4] He devised and implemented a course which is described alongside other courses that he believes achieve this goal by N. J. Vye, V. R. Delelos, M. S. Burns, and J. D. Bransford. Teaching thinking and problem solving: Illustrations and issues in R. J. Sternberg and E. E. Smith (Eds.), (1988). *The Psychology of Human Thought*. New York, Cambridge University Press. Ref. [3] is essentially a handbook that can be used a background for such courses. Most recently he has been concerned with teaching for wisdom. Sternberg, R. J., Jarvin, L., and E. L. Grigorenko (2009). *Wisdom, Intelligence, Creativity and Success*. New York. Skyshore Publishing. 185

[5] Sternberg, R., Conway, B. E., Ketron, J. L., and M. Bernstein (1981). Peoples conceptions of intelligence. *Journal of Personality and Social Psychology*, 41, pp. 37–55. 186

[6] Green, S. (1990). Analysis of personal transferable skills requested by employers in graduate recruitment advertisements. Sheffield. Sheffield Personal Skills Unit, University of Sheffield. 188

[7] Cited in Heywood, J. (1994). *Enterprise Learning and its Assessment in Higher Education*. Technical Report no. 20. Sheffield, Employment Department. 189

[8] Youngman, M. B., Oxtoby, R., Monk, J. D., and J. Heywood (1978). *Analysing Jobs*. Aldershot, Gower Press. 189

[9] Trevelyan, J. (2014). *The Making of an Expert Engineer*. London, CRC Press (Taylor and Francis). 189, 201

[10] *loc. cit.* Ref. [3]. 189

[11] Freeman, J. and P. Byrne (1976). *The Assessment of General Practice*. 2nd ed., London, Society for Research into Higher Education. 191

[12] Weston, C. A. and P. A. Cranton (1986). Selecting Instructional Strategies. *Journal of Higher Education*, 57(3), pp. 259–288. 191, 194

[13] Goleman, D. (1994). *Emotional Intelligence. Why it Matters more than IQ*. New York, Bantam Books. 193

[14] Zirkel, S. (2000). Social intelligence: The development and maintenance of purposive behaviour in Chapter 1 Bar-On, R. and J. D. E. Parker (Eds.), *The Handbook of Emotional Intelligence*. San Fransisco, Jossey Bass. 194

[15] Hedlund, J. and R. J. Sternberg (2000). Too many intelligences? Integrating social, emotional and practical intelligence in Bar-On, R. and J. D. E. Parker (Eds.), *The Handbook of Emotional Intelligence*. San Fransisco, Jossey Bass. 194, 201

[16] *ibid* 195

[17] Bar-On, R. (2000). Emotional and social intelligence: Insights from emotional quotient inventory in Bar-On, R. and J. D. E. Parker (Eds.), *The Handbook of Emotional Intelligence*. San Fransisco, Jossey Bass. 195

[18] Culver, R. S. (1988). A review of emotional intelligence by Daniel Goldman. Implications for technical education. *ASEE/IEEE Proceedings Frontiers in Education Conference*, pp. 855–860. 195

[19] Stone, K. F. and H. Q. Dillehunt (1978). *Self Science. The Subject in Me*. Santa Monica, Goodyear Publications. 195

[20] *loc. cit.* Ref. [15]. 195

[21] Sternberg, R. J. and E. L. Grigorenko (2000). Practical intelligence and its development in Bar-On, R. and J. D. E. Parker (Eds.), *The Handbook of Emotional Intelligence*. San Fransisco, Jossey Bass. 195, 201

[22] Polyani, M. (1966). *The Tacit Dimension*. Garden City, Doubleday. 195

[23] Nouse alternative Nous. Intuitive apprehension: Common sense, practical intelligence, gumption. *The New Shorter Oxford English Dictionary*. Oxford University Press. 195

[24] *loc. cit.* ref. [9]. James Trevelyan writes that engineers should have "the ability to value, acquire, develop, and use tacit ingenuity which is compiled in a vast library in your mind composed of 'how to' fragments of unwritten technical and other knowledge. Your progress as a student depended on knowledge that you could write down in examinations, tests, quizzes, etc. In engineering, your progress depends much more on knowledge that is mostly unwritten, the kind that is carried out in your mind and the minds of other people. To acquire this knowledge, you may need to strengthen your ability to listen, read and see accurately." […] "You need to understand what engineering is, how it works and why it is valuable. Value is a multidimensional concept: Economic value, namely making money for yourself and others, is just one dimension. Other include caring for other people, social justice, sustainability, safety, social change, protecting the environment, security and defence […] you won't be able to find your way without knowing the point from which you are taking off. You will need the ability to understand yourself and where you are today" [pp. 43–44]. 195

[25] Giles Coren's column. *The Times*, April 22nd, 2017. 197

[26] *loc. cit.* Ref. [21]. 197

[27] Koort, B. and R. Reilly (2002). A pedagogical method for teaching scientific domain knowledge. *ASEE/IEEE Proceedings Frontiers in Education Conference*, 1, T3A-13 to 17. 197

J O U R N E Y   16

Social Reconstruction

16.1 THE FOURTH IDEOLOGY

The social reconstruction ideology takes the view that society is doomed because its institutions are incapable of solving the social problems, therefore education should reconstruct society. Schools and their curriculum have to be designed to achieve this purpose. This is in marked contrast to universities where little thought is given to the design of buildings for learning. The models they have inherited, and with which they appear to be satisfied, is a legacy of the industrial revolution.

Philosophically this ideology has its foundations in John Dewey's *Reconstruction in Philosophy* and *Democracy and Education* [1, 2]. As might be expected it is founded on a social constructivist view of knowledge—knowledge is relative. The purpose of teaching is to stimulate students to reconstruct themselves so that they can help reconstruct society. Some authors who could be associated with this ideology see teaching as a subversive activity [3].

Schiro writes "human experience, education truth and knowledge are socially defined. Human experience is believed to be fundamentally shaped by cultural factors; 'meaning' in people's lives is defined in terms of their relationship to society. Education is viewed as a function of the society that supports it and is defined in the context of a particular culture. Truth and knowledge are defined by cultural assumptions: they are idiosyncratic to each society and testable according to criteria based in social consensus rather than empiricism or logic" [4].

The principle methods of teaching are the "discussion" and "experience" group methods. In the "discussion" method the teacher elicits "from the students meanings that they have already stored up so that they may subject those meanings to a testing, verifying, reordering, reclassifying, modifying and extending process" [5]. In this way a transformation of and reconstruction of knowledge occurs in response to the group process. The experience method places "the students in an environment where they encounter a social crisis and learn from those who usually function in that environment" [6]. The teacher in this technique becomes colleague and friend.

This ideology may seem way outside the scope of engineering education, but is it? There are a number of reasons why the answer to the question is, "it is."

First, in respect of teaching method engineering educator Karl Smith has promoted "constructive controversy."

16.2   CONSTRUCTIVE CONTROVERSY

"Constructive controversy" exists when one person's ideas, information, conclusions, theories and opinions are incompatible with those of another and the two seek to reach an agreement [7]. Daniels and Cajander in Sweden used a staged approach as a guide to understanding how constructive controversy could be used as a scaffold. They followed the six stages of constructive controversy outlined by Johnson and Johnson which are:

1. Students are assigned problem/decision, initial conclusion.

2. Students present and listen and are confronted with an opposing position.

3. Students experience uncertainty, cognitive conflict, and disequilibrium.

4. Cooperative controversy.

5. Epistemic curiosity, information search.

6. Incorporation, adaptation to diverse perspectives, new conclusion.

Daniels and Cajander describe their experiences in two papers [8, 9]. They found that things did not always turn out as planned. For example, stage 4 overlapped with stage 2 in that controversy about how to cooperate took place in stage 2. In stage 5, while curiosity was sparked about what could make the project better, the students who had worked in sub-groups on parts of the project felt pressed to deliver their part of the project. There was evidence of insights and ideas, but they were not seen as something to act on, but as things to be noted. "There was not enough incentive to change what they were doing." Daniels and Cajander concluded that it was possible to use the model in a less structured manner. While it seems to have potential in educational settings, what can students learn from it that will help them in an industrial situation where there are problems of collaboration and conflict? ("How may constructive controversy be used to resolve conflicts within teams?").

Daniels and Cajander emphasize that when running courses in non-traditional settings it is necessary to explain the pedagogic reasoning for the choice of instructional method [10].

16.3   DEBATES

One thing that students might begin to learn is the art of communication. One method of developing communication skills is by debating. Alford and Surdu reported on their use in computer science at the U.S. Military Academy [11]. They argued that debates could:

*Help students organize and synthesize information (i.e., higher order thinking) and that the degree to which they do that, is similar to a "thorough end of term study for an examination."

*Encourage students to learn on their own.

*Increase student's cooperative skills.

*Improve verbal skills.

For a debate to be effective it requires a good topic. These may come from (1) topics that have been discussed in depth in the course, (2) topics discussed briefly during the course, and (3) relevant topics not discussed in the semester. The first encourages analysis and synthesis; the second encourages the development of the student's general knowledge; the third encourages the application of what is learned on the course.

There are several possibilities for assigning the position that students should take in the debate. For example, the students have to prepare half the topic and they are told which position they have to attend.

To ensure they learn both halves, they can be told to prepare for a debate without being told which position they have to take until much nearer the time of the debate. In this situation they are forced to evaluate the arguments for and against.

Another approach is to assign the roles at the beginning of the debate but this requires substantial preparation. A period of time has to be allowed at the beginning of the debate for the team to work out their strategy unless they are required to do this beforehand.

The United States Military Academy tried student vs. student, student vs. faculty, and faculty vs. faculty debates. The authors preferred the student vs. faculty debate because this produced a high level of effort from the students.

16.4   MOCK TRIALS

Mock trials have many similarities with debates. At the University of Valparaiso (Indiana) senior civil engineering students joined with third year law students to represent the plaintiffs, designer and contractors in each of three trials. The engineers were to be the expert witnesses, and they had to explain in lay terms all the technical concepts involved in the case, and also to give their opinion of the probable cause of failure. This involved them in several meetings with the law students who had to prepare the engineering students to act as expert witnesses. The cases were argued in front of a practicing judge.

Tarhini and Vandercoy [12] who conducted these classes reported that, "it forced students to completely understand the causes of structural collapse so they might clearly understand those causes, and it compelled collaboration with the third year students." They also noted that it provided an introduction to professional responsibility through the application of the ASCE Code of Ethics to the behavior of expert witnesses. That is, to render opinions based on facts only.

Inevitably the engineering students had to learn a new language and the logic of argument used by lawyers.

16.5   TURNING THE WORLD UPSIDE DOWN

It may be argued, that neither, debates or mock-trials lead to change in the way that the social reconstruction ideology envisages change. But change cannot be imposed it has to be argued for, a case has to be made. Debates and mock-trials help develop the skills of argument.

What matters is that engineering is by its very nature an agent of change. Engineering design is a social activity not only in its process an implementation [13], but in the impact that it has on society. Witness the enormous social change that is happening as a result of social media. Engineers do not seem have understood these consequences but to have supported these developments without questioning them as forces for both good and evil [14].

But engineers can be a force for good. Consider the work that has been done to promote peace engineering [15], or social justice [16]. In any event it is unlikely that engineering educators would disagree with the view that today the primary purpose of engineering is to improve the lot of individuals and the society in which they live. It reconstructs society, Social reconstruction educators take the view that "man is shaped by society and man can shape society [..] Individuals must first reconstruct themselves before they can reconstruct society" [17]. Engineering has a moral purpose but it originates in the individual.

The implications of social reconstruction ideology for the engineering curriculum and its teaching are, therefore, profound.

16.6   A CASE STUDY FOR CONCLUSION

The purpose of this book was to provide an introduction to teaching for beginning engineering educators and a means of reflection for experienced engineering educators. It began by showing that the problems beginning engineering educators have are the same problems that beginning teachers have. Moreover, a great deal of research and development has been completed on teacher education that is valuable to engineering education. An attempt has been made to demonstrate this point throughout the book. A major finding of that work is that if teachers are to experiment with instructional methods they have to receive the support of their colleagues.

It has been difficult to maintain this focus without veering into departmental and curriculum policies. Departments (schools) have to own the innovation even though they do not implement it. Mostly this does not happen in engineering so the teacher who wishes to innovate is on his/her own. But there is an awful lot that teachers can do that will have an impact on students, and it may not be what we think.

In the United States a small private liberal arts college has been the subject of a ten year study to find out "How College Works" [18]. The authors Daniel Chambliss and Christopher Takacs asked the question, "Can students get more out of college without spending more money?" This meant that they had to examine how the quality of college education could be improved without additional cost. The answer is surprising. They believed it could.

They found that the single most important thing in the quality of a student's education was to do with the way a college is organized to help the students with their relationships, and that went for the classroom experience as well. Relationships, "are the necessary precondition, the daily motivator, and the most valuable outcome. A student must have friends, needs good teachers, and benefits from mentors. A student must have friends, or she will drop out physically or withdraw mentally. When good teachers are encountered early, they legitimize academic involvement, while poor teachers destroy the reputation of departments and even entire institutions. Mentors, we found, can be valuable and even life changing......relationships are important because they raise or suppress the motivation to learn, a good college fosters the relationships that lead to motivation."

This text began with the idea that engineering educators become professional when they treat teaching as a scholarly activity, and the model of the teacher as researcher and or developer of his/her instruction was promoted. Since learning is shared activity the least an instructor can do to foster relationships is to share his/her scholarly activity with his/her students.

NOTES AND REFERENCES

[1] Dewey, J. (1948). *Reconstruction in Philosophy*. Boston, Beacon Press. 203

[2] Dewey, J. (1916). *Democracy and Education*. New York, Macmillan. 203

[3] Postman, N. and C. Weingartner (1969). *Teaching as a Subversive Activity*. New York, Delacorte Press. 203

[4] M. S. Schiro (2013). *Curriculum Theory. Conflicting Visions and Enduring Concerns*. 2nd ed., Los Angeles, Sage, p. 161. 203, 207, 208

[5] Postman and Weingartner cited by Schiro ref. [4]. 203

[6] *loc. cit.* Schiro ref. [4, p. 186]. 203

[7] Johnson, D. and R. Johnson (2007). *Creative Constructive Controversy Intellectual Challenge in the Classrooms*. 4th ed., Edina, MN. 204

[8] Daniels, M. and A. Cajende (2010). Experiences from using constructive controversy in an open ended group project. *ASEE/IEEE Proceedings Frontiers in Education Conference*. S3E-1 to 6. 204

[9] Laxer, C., Daniels, M., Cajander, A., and M. Wollowski (2009). Evolution of an international student project. *CRPIT Computing Education*, 95, pp. 111–118. 204

[10] Gorbet, R., Schoner, V., and G. Spencer (2008). Impact of learning transformation on performance in a cross-disciplinary project based course. *ASEE/IEEE Proceedings frontiers in Education Conference*, TC2-18 to 22. 204

[11] Alford, K. L. and J. R. Surdu (2002). Using in-class debates as teaching tool. *ASEE/IEEE Proceedings Frontiers in Education Conference*, S1F-10 to 15. 204

[12] Tarhini, K. M. and D. E. Vandercoy (2000). Engineering students as expert witnesses in mock trials. *ASEE/IEEE Proceedings Frontiers in Education Conference* 1, TIF-1 to 2. 205

[13] Bucciarelli, L. L. (2003). *Engineering Philosophy*. Delft, Delft University Press. 206

[14] Harari, Y. N. (2016). *Homo Deus: A Brief History of Tomorrow*. Harvill Secker. 206

[15] Vesilind, P. Arne (Ed.), (2005). *Peace Engineering. When Personal Values and Engineering Careers Converge*. Woodsville, NH, Lakeshire Press. 206

[16] Riley, D. (2008). *Engineering and Social Justice*. Morgan & Claypool. `http://www.morganclaypool.com` 206

[17] *loc. cit.* ref. [4, p. 163]. 206

[18] Chambliss, D. F. and C. G. Takacs (2014). *How College Works*. Cambridge, MA, Harvard University Press. 206

Author's Biography

JOHN HEYWOOD

John Heywood is a Professorial Fellow Emeritus of Trinity College Dublin-University of Dublin. He was given the best research publication award of the Division for the Professions of the American Educational Research Association for *Engineering Education: Research and Development in the Curriculum and Instruction* in 2006. Recently, he published *The Assessment of Learning in Engineering Education: Practice and Policy*. Previous studies among his 150 publications have included *Learning, Adaptability and Change; The Challenge for Education and Industry*, and co-authored *Analysing Jobs*, a study of engineers at work. He is a Fellow of the American Society for Engineering Education, a Fellow of the Institute of Electrical and Electronic Engineers, and an Honorary Fellow of the Institute of Engineers of Ireland. In 2016 he received the Pro Ecclesia et Pontifice Cross from the Pope for his services to education.

Author Index

Abercrombie, M. L. J., 13, 18
Alexander, C. J., 121
Alfey, H., 83
Alford, K. L., 204, 208
Anderson, L. W., 46, 51, 52
Andrews, F. M., 51, 52
Angelo, T. A., 33, 35
Apple, D. K., 61, 68
Aristotle, 99
Armstrong, T., 199
Atman, C. J., 69
Austin, G. A., 20
Ausubel, D. P. xvii, 123, 128, 131, 132

Badar, M. A., 77, 84
Baillie, C., 53, 147, 151
Ball, J., 142
Ball, K., 41, 44, 45
Barnes, L. B., 81
Bar-On, R., 195, 200, 201
Bassey, M., 4, 9
Batanov., D. M., 49, 54
Bates, M., 164, 170, 172
Bellon, E. C., 54, 129
Bellon, J. J., 54, 129
Berger, P., 96, 97
Berliner, D. C., 129
Bernstein, M., 200
Beston, S., 120
Bielefeldt, A., 184
Biggs, J. B., 53

Blair, A., 177
Blandin, B., 183, 184
Blank, M. A., 54, 129
Bloom, B. S., 19, 41, 42, 46, 47, 51, 52, 54
Blyth, W. A. L., 145, 151
Boffy, R., 91
Bolton, J., 60, 68
Boomer, G., 98
Borgford-Parnell, J., 69
Berglund, A., 165
Borglund, D., 164
Boykin, A. W., 107
Bransford, J. D., 150, 200
Brent, R., 166
Briggs, L., 162, 170, 171
Brophy, S. P., 55, 129, 134, 140
Brown, A. L., 129
Brown, P. C., 129, 165
Brownell, J. A., 107
Bruner, J., 19, 20, 88-90, 99, 102, 103, 105, 109, 115, 118, 123, 131
Bucciarelli, L. L., 134, 137, 208
Buchanan, W. W., 172
Burns, M. S., 15, 200
Burns, Robert, 11
Burt, C., 181
Butko, J. A., 69
Buttle, D., 166
Byrne, P., 191, 200

Caine, G., 20

Caine, R. N., 20
Cajander, A., 76, 83, 204, 207
Cameron, L. A., 168
Campbell, B., 199
Campbell, L., 199
Campione, J. C., 129
Cardella, M, 61, 68, 75
Carlsen, R. W., 106, 107
Carr, K., 35
Carroll, J., 174, 179
Carroll, P., 166, 167
Carter, G., 51, 68, 84, 85, 129
Carter, Sir Charles, 99
Casey, P., 179
Catalano, R. E., 165
Cattell, R. B., 179
Chambliss, D. F., 206, 208
Champagne, A. N., 137
Chen, A. S-Y., 83
Cheville, R. A., 4, 6, 9, 129
Chiu, C., 107
Cho, Y-H., 67
Chookittikul, W., 49, 54
Clement, J., 133, 134, 137
Cocchiarella, M. J., 150
Cohen, L., 47, 48, 54
Cohn, M. M, 1, 7
Collins, H., 4, 9
Collis, K. F., 53
Combs, A. W., 20
Conway, B. E., 200
Cooney, E., 83
Copleston, F., 103
Coren, G., 201
Corno, L., 128, 129
Courter, S., 8
Cowan, J., 54, 73, 77, 82, 118, 119, 122,
    134, 135, 141, 153, 164, 167
Cox, P., 172

Crain, W., 103
Cranton, P. A., 191, 194, 200
Crawford, W. R., 34, 135, 141
Cromwell, L. S., 58, 67
Cross, K. P., 17, 20, 33, 35
Crown, S., 150
Crynes, B. L., 7
Crynes, D. A., 7
Culver, R. S., 51, 114, 115, 120, 122, 172,
    195, 197, 201
Cuthbert, J. N., 172
Cyganski, D., 119

D'Amour, G., 81
Dai, X., 172
Daniels, M., 83, 204, 207, 208
Das, N., 55
Davidovic, A., 142
Davies, D., 113, 120
Davies, J. W., 164
Day, J. C., 129
De Cecco, J. P., 34, 135, 141
De Raad, B., 172
Dean, R. H., 60, 68
Deary, I. J., 174, 176, 179, 180
Dee, K. C., 165, 171
Delane, A., 179
Delatte, N., 129
Delelos,, V. R., 200
Dench, G., 177
Devitt, F., 56
Dewey, J., 118, 153, 201, 203, 207
DiBello, L.V., 134, 140
Diebel, K., 69
Dillehunt, H. Q., 201
Dimmit, N. J., 49
Dixon, N. M., 162, 166, 168
Doll, W. E., 118, 119
Donald, J. G., 149, 151

Donovan, I., 92, 93
Donovan, M. S., 150
Douglas, E., 106
Dressel, P, 67
Duncan-Hewitt, W. C., 120
Dunleavy, S. P., 143, 149

Eckerdal, A., 165
Edwards, R., 106, 134, 140
Eggleston, J., 87, 96, 97
Eisner, E., 12-14, 18, 43, 45, 52, 124
Elliott, J., 3, 8, 11, 17, 18, 21, 27, 34
Ellis, G., 165
Engler, B., 172
Ennis, R. H., 66
Entwistle, N. J., 82, 121
Epstein, R. M, 184
Erickson, H. L., 147, 151

Felder,G., 154-156
Felder, R. M., 161, 163, 166, 168
Fellows, S., 120
Fensham, P. J., 107
Ferro, P., 53
Festinger, L., 20
Feuerstein, D., 66
Figueiredo, J., 56, 69
Fitch, P., 115, 120, 122
FitzGibbon, A., 158, 168, 172
FitzGibbon, R. E., 78
Flammer, G. H., 120
Flinders, D. J., 119
Floud, J., 183
Fordyce, D., 66, 68
Frank, B. M., 83
Freeman, R., 150
Freeman, S., 81
Freeman, J., 166, 191, 200
French, W. L., 20

Froyd, J. E., 107
Fuentes, A., 150
Fuge, W., 97
Fuller, M., 71, 80

Gage, N. L., 129
Gagne, R., 57, 91, 123, 131, 132, 137, 145
Gardener, H., 175, 185, 186, 193, 198, 199
Garick, P., 137
Gehringer, E., 50, 55
Gibbs, G., 53
Gluesing, J., 69
Godwin, D. B., 150
Goleman, D., 193, 200
Goodenough, D. R., 170
Goodhew, P., 147
Goodnow, J. J., 20
Goold, E., 56
Gorbert, R., 207
Grasha, A., 156, 161, 163, 166, 170
Green, S., 200
Gregory, S. A., 52
Griffin, C., 182, 184
Grigorenko, E. L, 187, 189, 199-201
Grimson, W., 57, 67
Gunstone, R. E., 137

Hackos, J. T., 51, 114, 120
Hadow, Sir Henry, 178
Hall, S. R., 137
Halsey, A., 183
Hanley, M., 121
Hansen, M. A., 140
Harb, J. N., 168
Hargreaves, K., 119
Harris, Y. N., 208
Harry, Prince, 197
Hawkins, J. D., 165
Haynes, A., 83

Hedlund, J., 194, 195, 200
Hendriks, A. A. J., 172
Hernstein, R. J., 173, 176, 183
Hestenes, J., 140
Heywood, J., 4, 6, 8, 9, 18, 27, 29, 34, 67, 82, 98, 103, 128, 165, 166, 178, 184, 200
Hirst, P., 98, 99, 101, 102
Hjalmarson, M. A., 118, 122
Hoare, C., 121
Hoffman, M., 66
Hofster, W. K. B., 172
Holdhusen, M, 47, 54
Honey, P., 167, 168
Hounsell, D., 82, 121
Howard, R. W., 14, 16, 31, 34, 35, 143, 150
Hoyle, E., 4, 9
Hoyt, B., 61, 68
Hsieh, C., 134, 140
Hu, Helen Hu., 106
Huddleton, C., 129
Humble, W., 56
Hundert, E. M., 184
Hurst, G., 7

Imrie, B. W., 53
Itahashi-Campbell, R., 69

Jackson, P. W., 14
James-Byrnes, C., 47, 54
Jarvin, L., 200
Jensen, A. R., 184
Johnson, D., 204, 207
Johnson, G., 150
Johnson, R., 204, 207
Johri, A., 67, 107, 140
Jonassen, D. H., 57, 67
Jovanovic, V., 184
Jung, C., 161

Kahney, H., 60, 68
Kardos, G., 71, 80
Kasprzak, E., 106
Kast, F. E., 20
Kaufman, J. C., 187, 189, 199
Kaupp, J. A., 83
Kean, A., 134, 140
Kelly, D. T., 51, 68, 84, 85, 129
Keogh, B., 141
Ketron, J. L., 200
Ketteridge, S., 172
Kiersey, D., 164, 170, 172
King, A. B., 107
King, P. H., 65, 69
King, P. M., 116, 117, 122
Kinsinger, K. S., 184
Kitchener, K. S., 116, 117, 122
Kline, P., 164, 172
Koen, W. V., 59, 68
Kolb, D., 29, 35, 156-158, 166
Kolhberg, L., 98, 103
Komsky, B. R., 83
Koort,B., 197, 198, 201
Korte, R., 57, 67
Kottkamp, R. B., 1, 7, 22
Krathwohl, D. R., 46, 52
Krause, S., 145, 150
Krick, E. V., 29
Krishnan, M., 119
Kruisman, G., 172
Kussmaul, C. L., 106

Laxer, C., 207
Land, R., 147
Langer, E. J., 121
Larkin, J. H., 73, 82
Lee, W., 129
Lesser, G., 103
Liebold, B. G., 69

Liete, C., 83
Lin, Y., 77, 84
Lipman, M., 67, 89
Litzinger, T. A., 122
Livesay, G., 165
Lo, J., 120
Lohmann, J. R., 107
Lortie, D. C., 1, 7
Lubkin, J., 80
Luchins, A. S., 20, 82
Luckman, T., 96, 97
Lydon, P., 73, 82
Litynski, D. M., 68

Maarek, J-M., 106
Macfarlane Smith, I., 178
Macmurray, J., 51, 53, 103
Mager, R. F., 47
Mak, F.,129
Mallikarjunan, K., 119, 120
Mann, G., 129
Mannion, L., 47, 48, 54
Marra, R. M., 122
Marshall, S., 19, 20
Martin, F., 183
Marton, F., 82, 121
Maslow, A., 71, 81
Matthews, G. B., 89, 105
Mayfield, C., 106
McCarthy, B., 161, 168
McCauley, M., 162, 172
McComisky, J. G., 166
McCracken, W. M., 75, 82
McDaniel, M. A., 129, 165
McDonald, F. J., 20, 52, 58, 67, 82, 129, 165
McGuinness, S., 69
McKernan, J., 35
McNally, H., 171
McPeck, J., 66

Mentkowski, M., 122
Meyer, J. H. F., 147
Miertschein, S., 151
Miller, R. I., 50, 56
Miller, Ron., 134, 140, 143, 150
Miller, R., 66
Mina, M., 155
Miwa, T., 120
Monk, J. D., 184, 200
Montagu Pollock, H., 18
Moore, T. J., 118, 122
Moreland, J. L. C., 69
Moriarty, M., 165
Morris, R., 106
Mujika, M. G., 69
Mumford, A., 167, 168
Murphy, D. E., 69
Murray, C., 173, 176, 183
Myers, L., 161, 162, 170, 171

Nair, J., 69
Nardi, A. H., 81
Nasr, R., 137
Naylor, S., 141
Nelson, M. A., 150
Newman, J. H., 17, 18
Newstetter, W., 75, 82
Nicolleti, D., 119
Noguera, P., 107
Norris, S. P., 66
Norwood, Sir Cyril., 177
Novak, J. D., 150
Nygren, K. P., 68

Olds, B. M., 67, 107, 140, 150
Orr, J. A., 119
Orton, A., 124, 128
Oscanayan, F. S, 67
Oswald, N., 129

Otter, S., 51
Owens, R. C., 3, 9
Owens, S., 83
Oxtoby, R., 184, 200
Oyamo, C., 172

Palmer, B., 122
Papadopoulos, C., 147, 151
Parent, D., 172
Park, O., 150
Parker, J-D. E., 200
Parsons, J. R., 172
Pascarella, E. T., 77, 83, 122
Patrick, K., 142
Pears, A. N., 2, 8, 155, 165
Pellegrino, J. W., 134, 140, 150
Pelz, D. C., 51, 52
Perry, W., 4, 9, 51, 114-116, 121, 156
Perzer, S. Y., 69
Peters, A-K., 155, 165
Phenix, P. H., 98-103
Phillips, D. K., 35
Piaget, J., 87, 98, 104, 109, 114, 118, 157
Plants, H. L., 60, 68
Plato, 99
Plomin, R., 181, 184
Plowden, Lady, 153, 164
Plunkett, G., 148
Polya, G., 71, 72, 80
Polyani, M., 195, 201
Postman, N., 207
Prawat, R. S., 125, 129
Price, G., 76, 133
Prince, M., 61, 68
Prior, P, 21, 22

Radcliffe, D., 52
Raghavan, J., 144, 150
Ramsden, P., 53, 121

Rand, Y., 66
Ratcliffe, G., 121
Recktenwald, G., 106, 134, 140
Red, W. E., 57, 60, 67, 71, 80
Reilly, R., 197, 198, 201
Rhee, J., 172
Riley, D., 208
Rodriguez, L., 47
Rodriguez-Marek, E, 51
Roediger, H. L., 129, 165
Rokeach, M., 11, 17
Rosati, P. A., 71, 80
Rosch, E., 150
Rose, A., 184
Rosenzweig, J. E., 20
Ross, D. C., 69
Ross, J., 106
Ross, R., 106
Ross, S., 60, 68
Roth, W. M., 113, 120
Ruskin, A. M., 68
Rutherford, U., 164
Ruthven, K., 96-98
Rutshon, J., 172
Ryan, A., 110

Säljö, R., 82, 121
Sandberg, J., 183, 184
Saupe, J., 58, 67
Scandura, J. M., 124, 128
Schiro, M., 87, 88, 96, 98, 104107, 155, 164, 201, 207
Schmidt, H. G., 172
Schoner, V., 207
Schrödinger, E., 109, 110
Schubert, W. H., 20, 89, 90, 105
Schwartz, J. C., 137
Sgro, S., 81
Shahhosseini, A. M., 77, 84

Sharo, M., 172
Sharp, A. M., 67
Sharp, J. E., 168
Sherlock Holmes, 72
Shulman, L. S., 40, 90, 91, 105, 123, 128
Silverman, L. K., 161, 163, 168
Simon, H., 60
Svnicki, M., 162, 166, 168
Skemp, R. R., 134, 135, 140
Skryabina, E., 147
Smith, K., 57, 67, 203, 207
Smith, H. W., 98
Snyder, M. E., 179
Snygg, D., 20
Sockett, H., 1, 7
Soloman, B. S., 171
Sorby, P., 179
Sosniak, L. A., 51, 52
Spearman, C., 174, 181
Spencer, G., 207
Spenker, M., 179
Spens, Sir William., 177
Spezia, C., 129
Stager, R. A., 71-74, 81
Stannard, R., 89, 105
Steadman, M. H., 33, 35
Steif, P., 134, 140, 146, 149, 151
Stein, B., 83
Sternberg, R. J., 185-187, 189, 193, 195, 199-201
Stice, J. E., 47, 48, 168
Stone, K. F., 201
Strait, R., 120
Streveler, R. A., 134, 140, 150
Strobel, J., 61, 68, 75
Surdu, J. R., 203, 208
Sutinen, A., 59, 67
Svinicki, M. D., 162, 166, 168
Swackhamer, G., 140

Taba, H.,, 145, 151
Takacs, C., 206, 208
Tarhini, K. M., 205, 207, 208
Tarhio, J., 58, 67
Tasooji, A., 145, 150
Tennyson, R. D., 150
Terenzini, P. T., 77, 83, 122
Terman, L. M., 176
Terry, R. E., 168
Thompson, N., 60, 68
Thornton, S. J., 105, 119
Thouless, P. H., 20
Thurstone, L. L., 174, 178
Todd, R. H., 162, 168
Tomovic, M., 184
Trevelyan, J., 17, 20, 31, 35, 41, 50, 52, 56, 69, 183, 184, 195, 200, 201
Trichina, E., 142
Turiel, E., 103
Tyler, R., 39, 40, 51, 118, 119

Uria, E. S., 69

Vable, M., 137
Van der Molen, H. T., 172
van Helden, H. J., 29, 34
Vandercoy, D. E., 205, 208
Vernon, P. E., 51, 174, 178, 181, 184
Vesilind, P. A., 208
Vidic, A. D., 67
Vonk, J. H. C., 29, 34
Vye, N. J., 200

Wales, C. A., 71-74, 81
Walker, J. M. T., 65, 69
Walker, P., 53
Walther, J., 52
Ware, R., 172
Warren, J., 142
Wechsler, D., 174, 179
Weingartner, C., 207

Wells, J. M., 124, 128
Wells, M., 140
Wenger, F., 165
Weston, C. A., 191, 200
Wexler, C., 67
White, A., 165
Whitehead, A. N., 118
Whitfield, P. R., 156, 157, 166
Whitfield, R. C., 84, 102-104, 105
Whysong, K., 120
Wigal, C. M., 165
William, D., 8
Williams, B., 56, 69
Williams, M. W., 68
Williams, R., 88, 107
Willis, C., 151
Wilson, C., 141

Wirszup, I., 120
Witkin, H. A., 161, 170
Wittrock, M. C., 90, 91
Wollowski, M., 207
Womenldorf, C., 134, 140
Woods, D. R., 57, 67, 79, 80, 109, 115, 122
Woods, R. C., 120

Yao, D., 172
Yokomoto, C., 163, 172
Yost, S., 119
Young, M., 173, 177
Young, T., 177
Youngman, M. B., 184, 200

Zhang, D., 172
Zirkel, S., 200

Subject Index

ABET, 1, 3, 51, 154

Abstraction (levels of), xvii, 72, 98, 104, 109, 157

Academic/vocational divide, 174

Accommodators, 160, 161

Accountability *see* Journey 1

Accountability, 11, 18, 112, 145, 146

Action Research, 20-22, 27-30, 34, 35

Advanced Organiser, 123, 128, 145

Affective domain, 50, 53, 66, 191, 194, 195

Alignment, 3, 64

Analogy, 16, 144

Animation, 136, 141

Answerability, 1

Applied Intelligence, 187, 195, 200

Applied motivation, 120

Assessment, 3, 13, 31, 45, 51, 59, 61-64, 75-79, 84, 116, 121, 134, 141, 146, 151, 154-156, 175, 182, 187-189, 191

Assimilators, 158, 160, 161, 166

Attitudes, 2, 4, 16, 19, 22, 37, 40, 64, 87, 110, 113-115, 132, 148, 163, 182

Beyond IQ - A Triarchic View of Intelligence, 185-187

Bell curve controversy, 176

Big 5 Personality Inventory, 164

California Critical Thinking Skills test, 83

Children's discourse (about engineering), 110-112

Code (of ethics), 4, 6, 205

Cognitive development, 114-118

Cognitive dissonance, 17, 19

Cognitive domain, 41, 42, 50, 194

Cognitive organisation (structure), 66, 125, 128

Cognitive process, 43, 46

College of Preceptors, 25

Collegiate Learning Assessment, 83

Communication, 83, 113, 147, 182, 188, 192, 196, 197, 204

Community of care, 154

Community of practice, 154

Competence, xviii, 8, 47, 100, 109, 112, 188, 173, 182, 183, 185

Competence
    accidental, 52
    social, 185
    inside, 182
    outside, 182

Competency, 11, 13, 50, 59, 173, 176, 181-183

Complexity (levels of), 41, 60, 144, 178

Concept
    cartoons, 136, 141
    clusters, 140, 146, 149
    inventories, 10, 134, 140
    inventory, 145
    learning *see* Journey 10
    learning, 14, 31, 127, 144, 149
    maps, 16, 65, 142, 145, 146

of stress (mechanical), 44
Concepts
    complex, 132, 135, 143, 144
    fuzzy, 143, 144
Constructive Controversy 203, 204
Constructivist (view of knowledge), xix, 96,
    153, 205
Content
    syllabus, xv, 27, 39, 40, 48, 62, 64, 92,
        94, 98, 99, 121, 128, 138, 145, 146,
        151, 154, 159
    Gardener's, 175, 185, 186
Convergent thinking, 156, 157, 168
Cornell Test of Critical Thinking, 83
Critical thinking, 43, 45, 47, 57-59, 66, 71,
    76-78, 114-116, 189, 192, 193
Curriculum
    and cognitive development, 144-146
    paradigms (received, reflexive,
        restructuring), xvi, 2, 87, 96, 97,
        102
    process (complexity of), xvi, 62, 64, 153

Debates (learning), 191, 204, 205
Decision making (skills), 15, 26, 50, 59, 71,
    112, 157, 196
Deep learning, 53, 75, 109, 116, 121, 134,
    142, 146, 151
Democracy and Education, 203
Design (engineering), 41, 44, 45, 50, 61, 63,
    66, 71, 90, 91, 96, 113, 178, 179,
    206
Diagnosis (skill of), 50, 60, 77
Disciplines of knowledge (philosophical
    debate), xvii, 58, 87-89, 96, 97,
    100-102, 106, 115
Discovery (Inquiry) learning, xvii, 4, 26, 30,
    40, 64, 89-94, 99, 106, 115, 123,
    169

Divergent thinking, xvii, 156, 157, 166, 168

Educational connoisseurship, xv, 13, 14, 17,
    29
Educational decisions, 78, 98
Engineering educators - Beginning, xv, 2, 7,
    47, 89
Engineering Science (A level), 41, 62, 63,
    76, 84, 90, 93, 95, 96, 124
Enterprise in Higher Education Initiative,
    189
Evaluation, 2, 6, 13, 14, 16, 17, 26-33, 40,
    59, 60, 84, 112, 119, 134, 140, 142,
    144, 161, 194
Examination(s), xvii, 41, 54, 56, 68, 125,
    155, 162, 173, 201
Examples, xvii, 14, 16, 26, 31, 33, 34, 94,
    134-136, 142, 144, 163
Experiential learning (Kolb's theory of), xviii,
    106, 157-159
Expert(s), xvi, 16, 17, 31, 41, 50, 61, 65, 66,
    73, 89, 145, 154, 158, 205
Expository instruction, xvii, 26, 30, 40, 64,
    91-94
Expressive activities, 45, 50, 52, 64

Fields of knowledge, 101, 102
Forms of knowledge, 53, 98, 99, 101, 102
Frames of Mind. The Theory of Multiple
    Intelligences, 175

General practice (assessment of), 191
Grading, 26
Guided design, xvi, 71, 72
Guided discovery, 10, 30, 32, 91-93, 169

Heuristic(s), xvi, 14, 15, 19, 26, 59, 65,
    71-74, 118, 140
Higher Order Thinking Skills (HOTS), 43,
    48, 80, 131, 132, 204

Honey and Mumford Learning Styles Inventory, 167

Immersion (critical thinking), 59, 66
Independent learning, 113
Index of Learning Styles, 156, 161, 162
Industry (role in development), 182, 183
Inquiry learning *see* discovery
Instruction, xvi, xvii, xviii, 1, 12, 25, 28, 40, 52, 62, 64, 77, 80, 82, 87, 96, 99, 109, 121, 123-126, 132, 134, 136, 140, 144, 145, 158, 164
Instructional planning (design), 7, 17, 22, 26, 27, 29, 47, 48-50, 74, 89, 157, 204, 206
Intellectual development
    Perry model, xvii, 4, 114-117, 156
    King and Kitchener's model, xvii, 114, 116, 117
Intellectual Development *see* Journey 8
Intellectual Development, xvii, 4, 156
Intelligence, 14, 104, 173-180
    and employment, 175, 176
    and experience, 185, 187
    and social class, 177, 181
    implicit theories of, 185-188
    Cattell's definition, 179
    Vernon's definition, 174, 178, 181
Interdisciplinary, 72, 173, 185

Joint Matriculation Board, 43, 63, 84, 95

K-12, 1, 66, 114, 127
Key Concepts, 145, 147, 148, 151
Kinesthetic activities, 154, 186
Knowledge, xv, xvi, xvii, 3, 8, 12, 13, 17-19, 25, 26, 30, 39, 42, 46, 57, 61, 66, 72, 87, 88, 96-102, 107, 123, 124, 126, 146, 151, 153, 155, 161, 165, 182, 187, 201, 203, 205.

tacit knowledge, 13, 14, 17, 25, 26, 30, 53, 195-197, 201, 203, 205

Language(s) (of engineering), xvi, 75, 113, 205
Learner Centred Ideology, xviii, 153-155
Learner (s), xvii, 12, 31, 39, 47, 53, 57, 59, 89, 90, 95, 99, 113, 123, 124, 124, 128, 131, 132, 134, 144, 153, 158, 161, 197, 199
Learning
    levels of, xvii, 53, 62, 65, 72, 80, 131, 194
    centred ideology, xvii, 95, 153, 176, 183
    perceptual, 12, 13, 19, 178
    styles, xviii, 26, 155-161, 166, 167, 197
    styles inventory (Kolb), xviii, 29, 156-158, 161, 166, 167, 197
Learning-to-learn, 191
Lectures (lessons), xv, 11, 12, 25, 26, 28, 75, 90, 115, 123, 124, 145
Legacy Learning Cycle, 145
Lesson planning, 16, 26, 28, 34, 47, 126
Listening, 12, 31, 189

Man a Course of Study (MACOS), 89, 118
Maslow's Hierarchy of Needs, 71, 81
Materials Concept Inventory, 145
Mathematics (cultural interpretation), 96
McMaster Problem Solving Course in Engineering, xvi, 57, 71, 79
Mediators, 136
Memory, xvii, 19, 48, 113, 125, 178, 179
Metaphor, 139, 144, 182
Microteaching, 12
Middle schools, 145
Midwest Coalition for Comprehensive Design education, 183
Mind Map *see* concept maps

Misperception, 133, 134
Mock trials, 205
Model - Eliciting Activities (MEA), 77, 134
Montessori Schools, 153
Motivation, 45, 65, 71, 81, 95, 99, 114, 121,
        133, 162, 188, 207
Myers Briggs Temperament Indicator
        (MBTI), 161-164

Naïve knowledge, 134
National Society for the Study of Education,
        42
Nature, 142, 176, 181
Negotiate (ion), 15, 31, 96, 114, 115
Novice(s), xvi, 60, 65, 66, 73, 75, 145, 167
Nurture, 173, 176, 181, 182

Objectives
        behavioural, 13, 17, 39-43, 47-50,
                62-64, 65-77, 104, 124, 131, 145,
                153
        focussing, 50, 56
        movement, xvi, 13, 39, 43, 50, 79
Organisation for learning see Journey 9
Originality, 41, 51
Outcomes, 39, 41-45, 47, 50

Paul Elder model (critical thinking), 83
Peer
        review, 2, 154
        teaching, 154, 194
Personal transferable skills, 100, 189, 195
Piaget see stages of development
Plowden report, 153
Polya's model of decision making, 71
Practical reflection, 3, 17
Primary Mental Ability Test, 178
Principles, 44, 54, 64, 72, 90, 94, 99,
        123-126, 131, 134, 141, 154, 161
Prior knowledge, xvii, 123, 124, 125-127

Problem Based learning (PBL), xvi, 57, 61
Problem
        finding, 50, 59
        management see Journey 5
        management, 59
        solving see Journey 5 and 6
        solving, 14, 19, 26, 43, 47, 50, 51, 54, 57
                61, 79, 80-82, 104, 110, 117, 132,
                133, 156, 157, 161, 178, 188, 195
        solving (open ended), 60, 72, 162
Problems
        closed, 60, 72
        open, 60, 72
        real world, 61, 77, 185
Procedural knowledge, 46, 147
Professional development (programmes),
        xiii, xv, 6, 18, 29
Professionalism- (–extended, -restricted), xv,
        4, 5
Project
        management, 31, 183
        planning, 62, 63
Prototype (example), 144
Psychometric testing, xviii, 154, 164, 174,
        175, 195
Psychomotor domain, 41, 191, 195

Qualitative understanding, xvi, 33, 73, 75,
        76, 80, 133, 134
Question design, 49, 53, 54, 64
Questions(ing), 48, 49, 115, 118, 145, 154,
        155, 161-163, 185

Recursive curriculum, 109, 110, 118
Reflection, 3, 14, 17, 21, 26, 30, 31, 115,
        119, 157, 161, 195, 206
Reflective Judgment Index, 116, 117
Reflective thinking, 14, 109, 116-118, 193
Reggio Emilia Schools, 153
Repetition, 110, 118, 167

Responsibility, xv, xviii, 1, 3, 7, 41, 116, 183, 205

Reverse engineering, 81

Salford University, 163
Scholar Academic Ideology *see* Journey 7
Scholar Academic Ideology, xvi, 57, 80, 182
Scholarship of teaching, xv, 17, 26, 33
Schrödinger's equation, 110
SCOOPE project, 110, 113, 114
Self-accountability, 7, 11, 12, 145
Self-assess(ment), 3, 154
Self-paced proctorial system, 121
Semantic Map, 144
Set mechanisation (induction), 82
Smith College, 154
Social efficiency ideology *see* Journey 4
Social efficiency ideology, xvi, 57, 79, 87, 153
Social reconstruction Ideology *see* Journey 16
Social reconstruction Ideology, xviii, 175, 176
Spatial ability, 174, 178, 179
Spiral curriculum, xvii, 53, 87, 90, 98, 104, 109, 114, 199
Stages of development (Piaget), xvii, 53, 87, 90, 104, 109, 114, 118, 157, 199
Standardised tests, 76, 154
Standards, 147, 151
Stanford Binet Intelligence Scale, 176
Surface learning, 116, 121

Synthesis, 41-44, 61, 85, 171, 205

Tacit ingenuity, 201
Tacit knowledge, 13, 14, 17, 25, 26, 30, 53, 195, 197
Teaching as research, 25, 33
Technical coordination, 31
Technique, 4, 7, 11, 41, 101, 124
Temperament and learning styles, 162, 163
The Bell Curve, 173, 176, 183
The New Republic, 176, 183
The Rise of the Meritocracy, 173, 177
The Taxonomy of Educational Objectives, xvi, 40-42, 61, 79, 100
Three Stratum Model, 174
Tipperary Leader Group, 110
Transfer (of knowledge, skill), 17, 57, 61, 62, 76, 93, 124, 125, 132, 144, 145

University of Sheffield Personal Transferable Skills Unit, 100, 189, 195
Uppsala University, xv, 8, 76
US Military Academy, 204

Visualisation, 41, 43, 66, 179

Watson Glaser Test of Critical Thinking, 82
Wechsler Adult Intelligence test, 174, 179
Wicked problems, 48, 60, 114
Woods Hole Conference, 89
Work, core competence, 183

Printed in the United States
by Baker & Taylor Publisher Services